SCALABLE MICROCHIP ION TRAPS
FOR QUANTUM COMPUTATION

DISSERTATION

zur Erlangung des
Doktorgrades Dr. rer. nat.
der Fakultät für Naturwissenschaften
der Universität Ulm

vorgelegt von

Stephan A. Schulz
aus Bremen

Amtierender Dekan	Prof. Dr. Peter Bäuerle
Erstgutachter	Prof. Dr. Ferdinand Schmidt-Kaler
Zweitgutachter	Prof. Dr. Tommaso Calarco
externer Gutachter	Prof. Dr. Ernst Rasel
Tag der Promotion	11.05.2009

The work described in this thesis was carried out at the

Universität Ulm
Institut für Quanteninformationsverarbeitung
Albert-Einstein-Allee 11
D-89069 Ulm
Germany

The experiment was funded by grants from the European networks
MICROTRAP and SCALA, the DFG in the framework of the
SFB/TRR21 and the Landesstiftung Baden-Württemberg.

When you can measure
what you are speaking about
and express it in numbers,
you know something about it;

but when you cannot express
it in numbers, your knowledge
is of a meagre and
unsatisfactory kind;

it may be the beginning of knowledge,
but you have scarcely in your thoughts
advanced to the state of science,
whatever the matter may be.

William Thompson, Lord Kelvin

Zusammenfassung

Die Entwicklung der experimentellen Quanteninformationsverarbeitung im Laufe der letzten Jahre zeigt, daß atomare Ionen gefangen in linearen Paulfallen ein bedeutendes Modellsystem zur Untersuchung von Quantenalgorithmen sind. Im Vergleich zu rein elektronischen Festkörpersystemen wie SQUIDs und Josephson-Kontakten bzw. der Kernspinresonanz bei Molekülen sind die gefangenen Ionen als Qubits weitgehend isoliert von störenden Einflussen aus der Umgebung. Die Verschränkung eines linearen Kristalls aus acht Ionen und die Verwirklichung verschiedener Quantengatter zeigen die auf der geringen Dekohärenz beruhende experimentelle Qualität des Modellsystems als Referenz. Die Weiterentwicklung der Quantenalgorithmen erfordert jedoch eine größere Zahl an beteiligten Qubits und damit höhere Ansprüche an die Skalierbarkeit linearer Ionenfallen. Die Herstellung makroskopischer linearer Ionenfallen wird durch Verfahren aus der Mikrosystemtechnik ersetzt. Dieser interdisziplinäre Ansatz bestehend aus Mikrotechnologie und Quantenoptik erlaubt die Realisierung von skalierbaren Mikroionenfallen für die Quanteninformationsverarbeitung.

Die vorliegende Arbeit umfaßt die Themengebiete der Entwicklung und Herstellung von skalierbaren miniaturisierten Ionenfallen, die experimentelle Integration und Charakterisierung und die experimentelle Demonstration der Qubit-Quantendynamik innerhalb der Mikrofallen. Es werden zwei grundlegend verschiedene Fallendesigns vorgestellt, numerisch simuliert und gefertigt - (a) die lineare Mikrofalle mit 56 Kontrollsegmenten zum Verschieben einzelner Ionen wurde mittels Lasermikrobearbeitung gefertigt. Ausgehend von Vorexperimenten mit Mikropartikeln wurden die planaren Oberflächenfallen optimiert und (b) mit einer linearen Testfalle mit 20 Kontrollsegmenten das Funktionsprinzip für atomare Ionen bestätigt. Ein neues Design mit einer Y-förmigen Kreuzung ermöglicht Verschiebeoperationen mit insgesamt 55 Kontrollsegmenten - das Design, die lithographische Herstellung und erste Funktionstests sind hierfür abgeschlossen.

Die lineare Mikrofalle wird mittels Seitenbandspektroskopie charakterisiert - durch Seitenbandkühlung wurde ein einzelnes Ion in den Bewegungsgrundzustand gekühlt und die Heizrate der Falle gemessen. Verbesserungen der Dekohärenz durch Optimierungen werden durch Ramsey-Spektroskopie belegt. Es werden zwei verschiedene Qubit-Systeme für das $^{40}Ca^+$ Ion realisiert - einerseits wird ein optisches Quantenbit über einen metastabilen Zustand und den Grundzustand, andererseits ein Spin-Quantenbit über die Zeeman-Aufspaltung des Grundzustands mit einem Raman-Übergang präpariert. Erstmals werden Messungen zum Ionentransport mittels Seitenbandspektroskopie realisiert und die Mikrofalle in einem weiten räumlichen Bereich delokal charakterisiert.

Die lineare Mikrofalle ist das Referenzprojekt für die Forschung zu einem europäischen Beitrag zum Quantencomputer. Es ist die erste Entwicklung einer skalierbaren Mikroionenfalle in Europa - die Ergebnisse dieser Arbeit zeigen nicht nur das Design und die Herstellung, sonder darüberhinaus durch die experimentelle Charakterisierung, z.B. durch die Bestimmung der Heizrate, die Eignung der Mikroionenfalle für Experimente der Quanteninformationsverarbeitung. Diese nach 1.5 Jahren koninuierlichem Betrieb immer noch in Europa einzigartige skalierbare Mikroionenfalle wurde als Referenzprojekt anderen europäischen Forschergruppen als modularer Bausatz zur Verfügung gestellt. Das Design bildet die Grundlage für die weitere Entwicklung der skalierbaren dreidimensionalen Mikroionenfallen.

Als anderer Ansatz werden die planaren Oberflächenfallen auf der Grundlage der planaren Falle mit Mikropartikeln als Referenzdesign entwickelt. Die Y-förmige Geometrie zeigt die Skalierbarkeit des Fallendesigns für Qubit-Systeme einer größeren Anzahl von Ionen, es werden erstmals neben Verschiebe- auch Sortieroperationen bei planaren Fallen möglich sein. Anhand einer linearen Falle für atomare Ionen werden Parameter ermittelt und die Falle elementar charakterisiert. Das Design der Y-Falle beruht auf mathematischer Optimierung der numerisch ermittelten Felder, die Fertigung zeigt die Realisierbarkeit des Ansatzes. Nach ersten elementaren Funktionstests wird die Falle derzeit im UHV-Rezipienten getestet.

Die weitere technologische Entwicklung der skalierbaren Mikrofallen hat direkte Auswirkungen auf die Realisierung von skalierbaren Quantenalgorithmen mit einer großen Anzahl von Quantenbits. Verschiebeoperationen in Speicher- und Prozessorbereiche während der Ausführung von Algorithmen werden erstmals die Paulfalle als dynamisches Element integrieren und erfordern einen interdisziplinären Ansatz zur Optimierung der Dekohärenzprozesse. Desweiteren ist die Hybridisierung in der Mikrofallenentwicklung eine vielversprechende Komponente - die Integration optischer Fasern in den Fallenchip wird zur faserintegrierten Mikrochipfalle als optoelektronisches Bauelement führen und der Quanteninformationsverarbeitung fundamental neue Möglichkeiten eröffnen.

Journal Publications

The work described in this thesis is published partially in a number of journal articles:

- S. Schulz, U. Poschinger, F. Ziesel, and F. Schmidt-Kaler
  Sideband cooling and coherent dynamics in a microchip multi-segmented ion trap, New J. Phys. 10, 045007 (2008).

- S. Schulz und F. Schmidt-Kaler
  Quanteninformationsverarbeitung: Segmentierte Mikrochip-Falle für kalte Ionen, Physik in unserer Zeit 38, 162 (2007).

- S. Schulz, U. Poschinger, K. Singer, and F. Schmidt-Kaler
  Optimization of segmented linear Paul traps and transport of stored particles, Fortschr. Phys. 54, 648 (2006).

Additionally some technical development is published:

- S. Schulz and F. Schmidt-Kaler
  Unmagnetischer temperaturbeständiger Multi-Pin-Halter für Mikrochipfallen im Vakuum, Patent pending
  DE 102006023158 A1 (2006).

Further articles has been published during the work on this thesis in the framework of ion trapping and quantum optics:

- B. Zhao, S. Schulz, S. Meek, G. Meijer, and W. Schöllkopf
  Quantum reflection of helium atom beams from a microstructured grating, Phys. Rev. A 78, 010902(R) (2008).

- J. Meijer, T. Vogel, B. Burchard, I.W. Rangelow, L. Bischoff, J. Wrachtrup, M. Domhan, F. Jelezko, W. Schnitzler, S. Schulz, K. Singer, and F. Schmidt-Kaler
  Concept of deterministic single ion doping with sub-nm spatial resolution, Appl. Phys. A 83, 321 (2006).

Contents

Chapter 1

Introduction

In the last decade the research in the field of quantum information science was established as a building block in quantum optics by exceptional results towards the fundamentals of a future quantum computer. Theoretical proposals for some buidling blocks were followed by experimental realizations in the research fields of nuclear magnetic resonance spectroscopy, photonics, solid state science and spectroscopy on trapped ions. Especially the research on ion traps has stimulated the fields of quantum state engineering, quantum cryptography and quantum metrology strongly.

The first fundamental ideas for information processing based on quantum mechanics were introduced by Paul Benioff [Ben80, Ben82] and Richard Feynman [Fey82]. Further work on the computational models of a quantum computer by David Deutsch illustrates the approach of a quantum Turing machine [Deu85] and the quantum circuit model, which describes the computation by a finites set of quantum gates represented as unitary operations on a qubit register [Deu89]. The significance of a future quantum computer has to be defined by efficient quantum algorithms that outclasses the performance of classical computers. One of the first quantum algorithms was the more technical Deutsch-Josza algorithm, then algorithms for number factorization [Sho96] by Peter Shor and quantum search [Gro97] by Lov Grover were proposed. The Shor algorithm illustrates the significance of a quantum computer compared to classical computational methods, because the security of the public-key cryptography is based on the assumption of unsolvable large number factorization. The best classical algorithms scale exponentially with the number size, Shors algorithm shows a polynomial behaviour and a finite solution time seems possible. The applicability of the Shor algorithm on large numbers may question the durability of modern encryption techniques. As a consequence of the complex quantum algorithms, the number of the qubits increases and the decoherence processes limit the calculation efficiency. Using quantum error correction schemes [Ste96], quantum phase and bit-flip errors can be corrected [Sho95] using additional ancilla qubits.

A significant drawback was the technical demand of controlling large qubit systems. For example, a two-qubit system may be error corrected using two additional ancilla qubits [Ste97]. Large-scaled quantum algorithms based on a manifold of qubits require complete control of the qubit systems. An alternative approach is the technique of fault-tolerant quantum computing with an additional computational overhead, achieving a fault-tolerance at error rates on the order of 10^{-3} per single operation [Kni05, Rau07]. In summary, the scaling up from first proof-of-principle experiments to scalable quantum computing regarding the control of large qubit systems is a technical demanding challenging task.

Several physical systems are realized for the application of quantum information processing so far. A general characterization by means of some criteria [Div00] is given by David DiVincenzo, which each appropiate system fulfills:

- A scalable physical system with well-characterized qubits,

- the ability to initialize the state of the qubits to a simple well-defined state like $|00\ldots\rangle$,

- long relevant decoherence times compared to the gate operation time,

- a universal set of quantum gates, and

- a qubit specific measurement capability.

Extensive studies on the nuclear spins of molecules [Ram05, Van05], electron spins in quantum dots [Gor05, Hol04], flux qubits in Josephson junctions [Mcd05, Ste06b], polarization on photons [Kni01, Kok07] and internal states in neutral [Sch04b, Boy06, Ye08] and ionized atoms show the applicability.

Quantum algorithms are demonstrated experimentally using different qubit systems: The Deutsch-Josza algorithm [Lin98, Gul03], the Grover search algorithm [Jon98, Chu98, Ahn00, Wal05, Bri05] and the Shor algorithm [Van01] for the factorization of numbers. All quantum algorithms independent from the physical system are a sequence out of single qubit operations and quantum gates [Llo95], which are unitary transformations on specifically addressed qubits. The limitations on the physical systems [Ste98] are the number of controllable qubits and decoherence effects - especially the technique of ion trapping shows scalable trap designs for the handling of large qubit systems and provides good decoherence properties.

Atomic ions trapped in linear Paul traps represent an ideal isolated qubit system for the experimental realization and development of quantum information processing. Compared to the nuclear magnetic resonance experiments the linear ion traps combine an isolated qubit system with excellent decoherence properties and scalability of the trap design. Based on the proposal of Ignacio Cirac and Peter Zoller [Cir95], the fundamentals of ion trap quantum computers were investigated [Sor99, Ste00]. Each qubit is implemented by the internal states $|0\rangle$ and $|1\rangle$ of the atomic ion, the vibrational Coulomb interaction of ions stored in a linear configuration acts as the quantum bus. Lasers drive the single-qubit operations, the quantum gates are realized with lasers using the quantum bus, which is based on phonons [Sch03]. Several logic quantum gates are demonstrated so far - starting with the first two-qubit gate [Mon95b], a phase gate [Lei03b], a controlled-NOT gate, and a Molmer-Sorensen-type gate [Ben08a]. The entanglement of two [Tur98], four [Sac00], six [Lei05] and eight ions [Hae05] were demonstrated, generating the first quantum byte. The entanglement of two ions was used to demonstrate the fundamentals of quantum dense coding [Sch04a], entanglement purification [Rei06a] and process tomography [Roo04a]. The generation of a robust quantum memory in a decoherence-free subspace [Lan05] effects new perspectives for high-precision measurements using entanglement [Roo06]. A method for probabilistic entanglement allows long distance entanglements of two ions from different traps [Moe07]. Entanglement of three ions allows the generation of GHZ- and W-states [Roo04b] and was used for the first teleportation of a quantum state [Rie04, Bar04]. The principle of quantum error correction is demonstrated using a three-ion entanglement consisting out of two ancilla qubits and the target qubit [Chi04].

An alternative approach beside quantum information science with ion traps is the simulation of quantum systems [Por04a, Lam07]. Such an analogous quantum computer simulates the evolution of quantum spin Hamiltonians with a trapped linear ion crystal [Por04b, Fri08]. In a quantum simulator the evolution of the Hamiltonian is investigated directly from an initial state to the final state, contrary to an implementation of the Hamiltonian with a set of quantum gates.

The applicability of ion traps in quantum metrology experiments [Hal06] as atomic frequency standards [Ber98a, Bec01, Mar04] next to Cs-beam clocks, atomic fountain clocks [Jef02, Wil02], optical lattice clocks [Boy07] and a molecular clock [Ye01] demonstrate the outstanding stable operation and reproducibility of trapping single ions in the free space. The comparison of the optical frequency to the microwave based atomic clocks is realized using frequency chains [Ber99] or a frequency comb [Osk06]. The stability of fundamental constants are also tested using ion clocks [Biz03, Pei04, For07]. The resolution of high-precision frequency measurements is enhanced by using entangled states [Bol96, Lei04] for the measurement of atomic properties [Roo06] and further development of atomic clocks [Sch05, Ros07] is realized with an uncertainty on the order of 10^{-17} [Ros08].

Besides experiments on atomic ions using linear ion traps, i.e. investigating phase transitions of larger atomic crystals [Die87, Wak92, Rai92, Bow99, Bab02], techniques for sympathetical cooling of molecular ion crystals in li-

near ion traps were developed [Bly05] - leading to rovibrational spectroscopy
on single molecules [Rot06, Koe07], the storage of cold biomolecules [Off08]
and the investigation of single molecule chemical reactions [Hoj08]. The
applicability of molecular ions for high-precision measurements [Sch07] and
as single qubits like atomic ions is proposed so far [And06, Wal08, Dem02]
and may be demonstrated using microfabricated linear ion traps within the
next years.

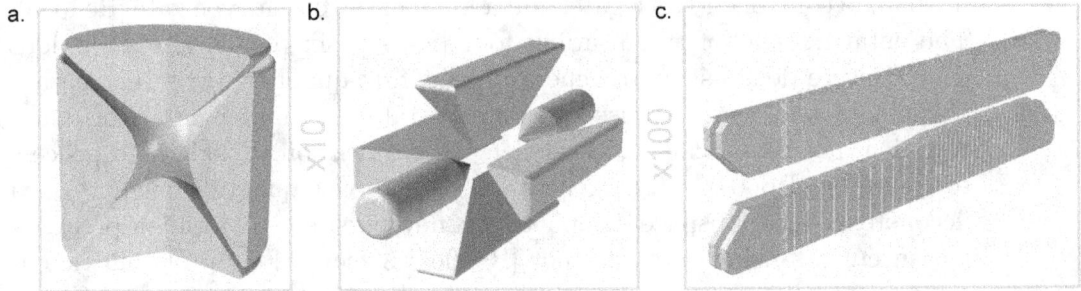

Figure 1.1: Evolution of Paul traps for quantum information experiments:
(a) The traditional Paul trap with macroscopic hyperbolic shaped electrodes
forms an ideal rf quadrupole potential. The two-dimensional dynamic confinement of the rf ring electrode (gray) is superimposed by the dc electrodes
(orange). Micromotion oscillations occur in all three directions in space.
(b) Linear macroscopic Paul traps benefit from the electrode design based
on a Paul mass filter. The micromotion is avoided perpendicular to the rf
confinement of the stored ions, a linear ion chain without any vibrational
excitation along this direction is trapped. Then the controlled excitation
of phonons along this trap axis is used as a quantum bus. (c) The microfabrication of linear Paul traps allows the miniaturization and segmentation
of the dc electrodes for the generation of overlapping micropotentials for
shuttling, splitting and merging of ion chains. Spatial separated zones for
the storage and processing show the advantages of scalable miniaturized
Paul trap development for quantum information science.

The successful development on the entanglement operations, quantum
algorithms and the minimization of decoherence effects over the decade illustrates the ability of ion trap quantum computing to scale up efficiently
the number of qubits for the realization of large scaled quantum operations [Kie92]. Apart from the more theoretical optimization of quantum
information processing, the environment, namely the ion traps (Fig. 1.1),
has to fullfil the requirements of scalability for the operation on large qubit
systems [Cir00]. A redesign of the traditional linear ion traps was unavoidable to meet the technical demands. Deduced from the Paul mass filter
[Pau53, Pau55], the linear traps designs are characterized by a large number
of segmentation electrodes in a three-dimensional or planar trap geometry [Chi05].

The implementation of microfabrication techniques allows the integration of a manifold of trap segments for the independent control of a large qubit system. The microfabrication of the sophisticated trap designs is the standard technique for scalable ion trap quantum computing today. Several three-dimensional designs are realized [Sch06, Hen06, Sti06]. The planar electrode traps are a different approach to a scalable trap design, and are characterized by a weaker trap depth, but better optical access and simpler fabrication methods [Sei06, Bro07, Lab08a, Bri06]. The heating rate as a fundamental trap property leads to the decoherence of the quantum state and is decreased by six orders of magnitude in a cryogenic environment [Lab08b]. The ability of the storage and control of a large qubit system requires the definition of several independent trap regions, connected by ion shuttling operations (Fig. 1.2). Highly segmented trap designs are privileged, because the speed for the quantum gates, or rather the phonon coupling of the quantum bus, is decreased with the number of ions confined in the same electric potential. Both areas in this field of research, the more theoretical investigation of quantum algorithms and the more practical development of scalable traps, meet each other at the realization of scalable quantum gates combining ion shuttling operations based on multi-segmented scalable microtraps characterized by low heating rates and decoherence.

In this thesis a new experiment is described mainly based on the design, development, fabrication and operation of a three-dimensional microtrap. Demonstrating coherent quantum state manipulation with ^{40}Ca$^+$ ions, Doppler cooling and sideband cooling to the motional ground state is realized. The heating rate is determined and the applicability for quantum computation is proven. Furthermore planar trap designs are investigated - a planar microparticle trap was built and operated. A linear microfabricated planar trap was operated with ^{40}Ca$^+$ ions, showing the principle of operation of a novel designed and fabricated Y-shaped planar trap. The thesis is structured as follows: Chapter (2) reviews the fundamentals of linear Paul traps, the principle of trapping, the stability diagram, and the differences between traditional Paul traps and microfabricated linear Paul traps are illustrated. The numerical optimization of the electric fields are investigated for the microchip trap, and ion shuttling operations in the non-adiabatic regime using optimal control methods show the operational fundamentals for modern quantum algorithms. Chapter (3) contains informations about the atomic properties of ^{40}Ca$^+$ and identifies the qubit schemes using the quadrupole transition for an optical qubit or Raman transitions for a spin qubit. Chapter (4) illustrates the concept of atom-light interactions and the laser cooling techniques like Doppler and sideband cooling to the motional ground state. Chapter (5) shows the fabrication techniques based on three different trap designs. The three-dimensional microchip trap is manufactured using laser structuring, the planar microparticle trap is produced

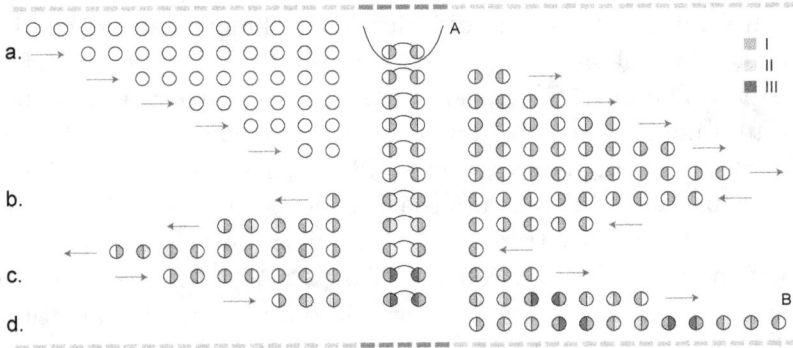

Figure 1.2: Entanglement scheme for multiple ions in a segmented micro-trap: The entanglement operation is based on a two-ion crystal in a spatial separated axial potential (A). In comparison to the multi-ion entanglement in a single potential, the entanglement operation is simplified because of the simple vibrational mode structure of the two-ion crystal. The speed of the entanglement operation is increased, additional time is used for the co-herent splitting, transporting and merging of the ions. The development of fast and efficient shuttling protocols preserving the coherence of the qubits is of broad interest. This entanglement scheme illustrates the segmentation of the electrodes in microtraps as a requirement for a scalable quantum processor: Starting with the entanglement (I) of 12 single ions to ion pairs (a), the subsequent entanglement (II) in groups of four ions (b) ends with the entanglement (III) of three groups of four ions (c). All of the 12 ions are entangled (B) by means of parallel shuttling and splitting operations. The laser interaction zone (dark gray) is enclosed by larger zones for transport and the storage (light gray).

using standard printed-circuit-board technology and the planar ion mi-crotraps are fabricated using photolithography techniques. Chapter (6) de-scribes the new experimental setup for the microchip trap and the planar ion microtrap, including solid-state laser systems, fluorescence detection optics, trap voltage supplies and experiment control system. Chapter (7) shows the experimental results regarding the microchip trap, including the coherent single ion dynamics, heating rate measurements, transport spectroscopy and quantum jump spectroscopy using Raman transitions. Chapter (8) describes the experimental results of the planar trap experiments. The microparticle trap operation is discussed, and shuttling and splitting operations are shown. The results of the planar ion trap experiments illustrates the axial confine-ments of an ion cloud in the linear ion trap. The installed Y-shaped planar trap is presented. Chapter (9) summarises the experimental results, chap-ter (10) illustrates future integration of fiber cavitites in the microchip trap. Finally, the appendix in chaper (11) shows layouts of the trap designs, a technique of rotational wafer coating and a design for a UHV compatible chip carrier socket supporting fast replacement of the microtraps.

Chapter 2

Fundamentals on scalable ion traps

The linear microtraps based on a Paul mass filter [Pau55] support especially the proposals for quantum information science by a scalable trap design and a dedicated direction of static confinement to establish the quantum bus. The micromotion is limited to the two-dimensional dynamic confinement in the radial cross section, the linear trap axis shows a pure static electric potential. The scalability is induced by microfabrication techniques to partition the linear trap axis with a manifold of static control electrodes. The optimization of the trap design and the operational conditions (2.1) are fundamental for the design of segmented linear Paul traps (2.2).

Scalable quantum algorithms profit from the design of the segmented linear microtraps. The quantum control of large linear ion crystals is simplified by splitting of the stored qubits. The quantum algorithms are realized on the subgroups separately, so the subsequent spatial transport of qubit information will be crucial for the scalable trap operation. The non-adiabatic single ion transport will be discussed and the constraints of motional heating during shuttling are investigated (2.5).

2.1   Linear Paul trap fundamentals

The first Paul trap in 1953 illustrates the mathematical fundamentals for ion trapping perfectly [Pau53, Pau90]. The hyperbolic-shaped ring combined with two endcaps reproduces the contour of an oscillating electric quadrupole potential - the oscillating electric fields are generated by the ring electrode, the endcap electrodes are supplied with a static potential.

a.   b.   c.   d.

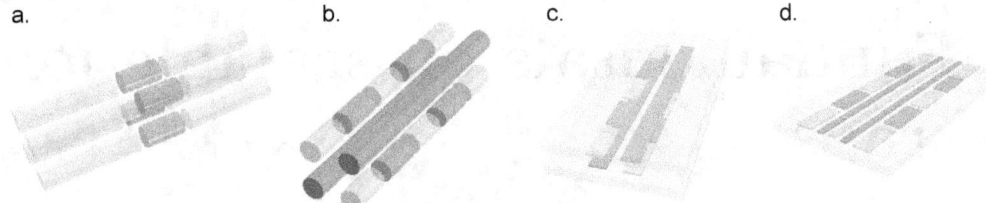

Figure 2.1: Linear segmented ion traps: (a) The single segment macroscopic ion trap is based on the standard Paul mass filter design. (b) The enhanced design with multiple segments illustrates the scheme of scalability. (c) The two-layer microchip design is adapted from the segmented macroscopic designs. (d) In the planar trap design all electrodes are located at a two-dimensional plane. The rf electrodes (blue) and the dc electrodes (red, gray) show the radial confinement and the static axial control of the trapped ions.

The miniaturization of ion traps started with designs of Paul-Straubel traps - a ring at an oscillating potential without endcaps, which are moved to infinity [Str55]. The simplified trap design was investigated explicitly in experiments [Deh67a, Yu91, Yu95] and is suitable for high-precision spectroscopy, i.e. ion clocks. The ring design is distinguished by the enhanced optical laser access and is scaled down easily, but shares the same disadvantages of the traditional Paul trap like micromotion in all directions.

In contrast to the three-dimensional traps the linear microtraps show a limited dynamical confinement to the radial cross section and the linear trap axis is idealized free of micromotion (Fig. 2.1). The trap design is based on the design of the Paul mass filter [Pau55]. The microfabrication techniques are utilized during the last decade for downscaling of the linear traps, while experimental experience on trapped ion control, i.e. Doppler and sideband cooling, was originated often by experiments using Paul-Straubel trap designs. The characteristics of linear trap operation and stability properties of trapping are borrowed from linear mass analysers.

2.1.1   Potential, stability regions and pseudopotential

The electric potential $\phi(x, y, z, t)$ for trapping of the ions is realized using a pure quadrupole potential $\phi_{tr}(x, y, z)$ consisting out of a dynamic potential and a static confinement represented by $\phi_{\text{dr}}(t)$ (Fig. 2.2). In a general approach the shape of the total electric potential given by the geometry of the trap electrodes is described by

$$\phi(x, y, z, t) = \frac{1}{2} \underbrace{\left(\alpha\, x^2 - \beta\, y^2 + \gamma\, z^2\right)}_{\phi_{\text{tr}}} \underbrace{\left(U_{\text{dc}} + U_{\text{rf}} \cdot \cos \Omega\, t\right)}_{\phi_{\text{dr}}}, \qquad (2.1)$$

with the trap drive frequency Ω and the trap drive voltage U_{rf} and U_{dc} for a static electric potential. The geometry factors α, β and γ are specified by the trap geometry and determined for each trap design numerically. The Laplace equation $\Delta\phi = 0$ causes the additional constraint for the geometry factors $\alpha - \beta + \gamma = 0$. The potential ϕ is valid for any three-dimensional or linear trap. Restricting to a linear trap means a two-dimensional dynamical confinement in x- and y-direction and a static potential at the linear trap axis, the oscillating electric potential ϕ is simplified to

$$\phi(x, y, t) = \frac{1}{2}\, \alpha\, (x^2 - y^2)\, (U_{\text{dc}} + U_{\text{rf}} \cdot \cos \Omega\, t), \qquad (2.2)$$

with a static potential for the confinement of the ions in z-direction

$$\phi_{\text{ax}}(z) = \frac{1}{2}\, \beta\, z^2 \cdot U_{\text{ax}}. \qquad (2.3)$$

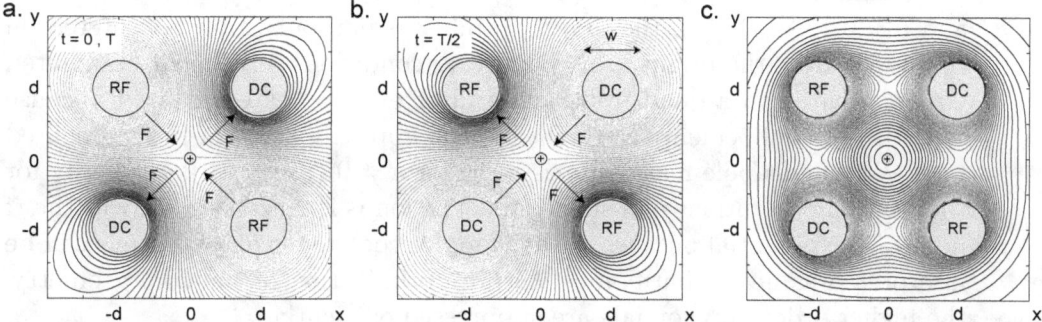

Figure 2.2: Electric potential and pseudopotential of a linear trap: The oscillating quadrupole potential of a linear rod-shaped trap is shown for $t = 0$ (a) and $t = T/2$ (b). The stable/unstable dynamical confinement of a single ion is alternating parallel to the principal axes of motion. The equipotential lines illustrate the high (orange) and low (gray) potential values. (c) The pseudopotential defined by the ponderomotive approximation shows the effective time-averaged harmonic potential.

The total electric potential for the linear trap results in the sum of de-coupled dynamical and static confinement $\phi_{lin}(x, y, z, t) = \phi(x, y, t) + \phi_{ax}(z)$. To illustrate the oscillating quadrupole potential (Fig. 2.2a,b), the equations of motion, i.e. $\ddot{x} = -Ze/m \cdot \partial\phi(x, y, z)/\partial x$, result to

$$\ddot{x} = -\frac{Ze}{m}\left(U_{dc} + U_{rf} \cdot \cos\Omega t\right)\alpha x \tag{2.4}$$

$$\ddot{y} = +\frac{Ze}{m}\left(U_{dc} + U_{rf} \cdot \cos\Omega t\right)\alpha y \tag{2.5}$$

$$\ddot{z} = -\frac{Ze}{m}U_{ax} \cdot \beta z .$$

The solution of the ion with charge Z and mass m for the dynamically potential in the xy-plane can be obtained analytically by a substitution to Mathieu equations [Lei03a], which are solved by the Floquet differential equation:

$$\frac{d^2 x}{d\xi^2} + \left(a_x - 2q_x \cdot \cos(2\xi)\right)x = 0 \tag{2.6}$$

$$\xi = \frac{\Omega t}{2} , \quad a = \frac{4Ze\,U_{dc}\,\alpha}{m\,\Omega^2} , \quad q = \frac{2Ze\,U_{dc}\,\alpha}{m\,\Omega^2}$$

In contrast to the persistent harmonic confinement of the ion along the linear trap axis the stability of the radial motion depends on pairs of the so-called stability parameters a and q (Fig. 2.3b). The condition of stability are co-domains in the aq-plane, the traps discussed here are operated near $a \approx 0$ and $q = 0.1, \ldots, 0.3$[1] in the lowest stability region only, that is centered on the a-axis near the origin. The operating conditions of a linear segmented trap compared to a quadrupole mass filter at $a = 0$ are idealized because of the axial confinement. Nevertheless the approximation of the segmented trap as a quadrupole mass filter is valid for $a \neq 0$ (Fig. 2.3b), especially for the trapping conditions of a single ion. The ion is localized strongly at the rf node of the potential because of the Doppler cooling, so all potentials can be interpreted as nearly harmonic. With a optimization of the trap geometry, the higher order polynomials are suppressed efficiently.

The frequency spectrum of a single ion trajectory is composed out of the secular frequency ω_{rad} and the frequency of the micromotion (Fig. 2.3a) identical to the trap drive Ω [Gos95]. The secular frequency is approximated at $a \ll 1, q \ll 1$ to $\omega_{rad} \approx \sqrt{1 + q^2/2} \cdot \Omega/2$. The micromotion is counter-phase to the trap drive and shows a q/2-smaller amplitude relative to the secular motion. The lowest order approximation of the x-component, respec-

---

[1]Instabilities effecting on ion loss caused by higher-order multipoles of the electric potential appears at $q > 0.3$ mainly [Alh95].

tively analogous to the y-component, and the z-component of the trapped ions motion is represented by

$$x(t) \sim \cos(\omega_{\text{rad}} t) \cdot \left(1 - \frac{q}{2} \cos(\Omega t)\right) \tag{2.7}$$

$$z(t) \sim \cos(\omega_{\text{ax}} t) \,, \; \omega_{\text{ax}} = \sqrt{Ze/m \cdot U_{\text{ax}} \beta}$$

The time-averaged pseudopotential [Deh67a, Lei03a] neglecting the micromotion of the trapped ions illustrates the harmonicity of the dynamic confinement in the xy-plane (Fig. 2.2c) and is used for effective trap depth calculations - the pseudopotential approximation [Deh67a, Deh67b] is based on the operation of the segmented linear trap as a linear Paul mass filter without the influence of the static confinement, and determines the trap depth of the pure dynamical potential:

$$\bar{\phi}(x, y) = \frac{(Ze)^2}{4m \, \Omega^2} \cdot |\nabla \phi(x, y, t)|^2 \tag{2.8}$$

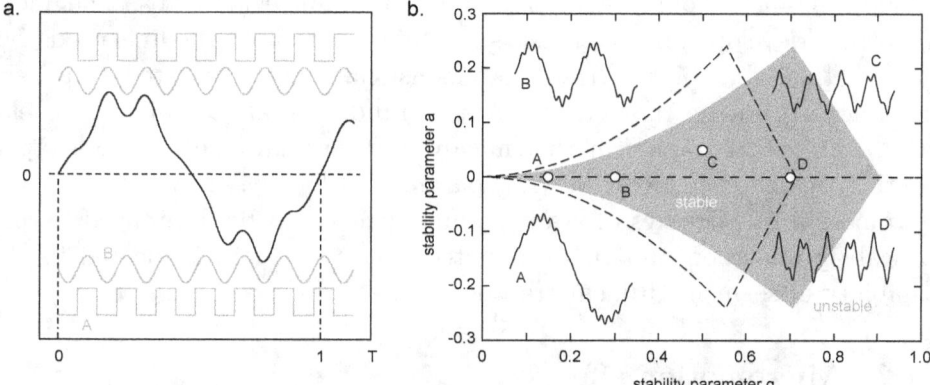

Figure 2.3: Ion trajectory and stability diagram of the linear ion trap: (a) The single ion trajectory shows the combination of secular motion (cycle duration T) and counter-phase micromotion. The microparticle trap is supplied with a rectangular-shaped trap drive (A), the ion traps with a sinusoidal waveform (A). (b) The lowest order stability diagram for a single ion is simulated numerically. The secular frequency is increased dependent on the q-parameter (A-D). The region of stability is calculated for a sinusoidal (orange) and a rectangular trap drive (blue). The rectangular stability diagram is squeezed in q-direction by a factor of $\pi/4$, which is the Fourier series prefactor of the trap drive voltage.

In a more accurate description of the trapping potentials and stability conditions of the segmented linear trap a full numerical approach is fundamental: The electric potentials deviates strongly from a pure quadrupole potential near the electrode surfaces, so a numerical simulation of the electric fields is preferred for the storage of linear ion crystals, especially for prolonged ion crystals in the radial direction like three-dimensional configurations. Higher multipoles of the electric field appears, i.e. the hexapole contribution $\phi \sim 3yx^2 - y^3$ [Alh95], to weaken the effective trap depth and engender determistic instabilites. This is relevant especially for the trapping of larger ion crytals.

At strong axial confinements resulting in a $\neq 0$, the influence of the control voltages at the different segments on the radial stability parameters has to be included. Further the inversed voltages on the control electrodes of a single pair for the compensation of micromotion lead to a linear electric field without potential offset at the location of the stored ion. The radial excursion of the stored single ion is enlarged.

The third unavoidable deviation from a pure linear trap configuration are components of the rf electric field from the trap drive parallel to the linear trap axis. Macroscopic linear trap designs consisting out of four linear rods without endcaps are the ideal case for neglecting the micromotion on the linear symmetry axis [Dre00]- therefore these trap designs are predestinated for experiments with large ion crystals, even because of their superior radial harmonicity compared to microfabricated linear ion traps. The linear microtraps developed in this thesis show micromotion on the linear axis because of their scalability - the two-dimensional microfabricated trap near the endcaps and the tapered region as a fundamental scalable design element. The planar traps near the endcaps and especially near the Y-shaped junction, which influences the trap parameters locally. In most cases the micromotion on the axes of static confinement could be minimized by numerical optimization and then neglected, but the origin is located in the scalability of segmented linear trap designs.

2.1.2 Micromotion effects

The micromotion of a single trapped ion is enlarged by a displacement of the ion out of the pseudopotential node (Fig. 2.4). Static or slowly varying electric potentials caused by field inhomogenities and patch charges lead to a position shift on the ion. Even asymmetries at the trap fabrication cause similar effects. Using a balanced voltage the ion is shifted towards the pseudopotential node and the micromotion is minimized (Fig. 2.4a).

The detection of the micromotion is realized for ion traps indirectly by the change in the fluorescence and directly by the measurement of the micromotion sideband amplitude using quantum jump spectroscopy. For the microparticle experiments using the planar trap, the detection is based on the time-averaged imaging of the ionized trapped microparticle with a CCD camera. The position of the microparticle is localized by the scattered light from a laser. The trajectory simulation illustrates the enlarged micromotion amplitude for a slight displacement out of the pseudopotential (Fig. 2.4b). The direction of the enlarged amplitude is directed to the shifted position. In the microparticle trap the time-averaged image of an unilateral uncompressed trajectory to the electrode surface confirms the numerical simulations exactly.

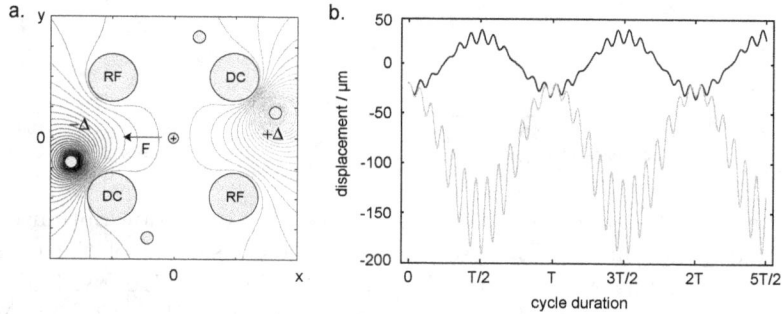

Figure 2.4: Enlarged trajectory due to micromotion: (a) Two pairs of compensation electrodes generate a homogeneous electric potential at the ions position in both directions. The ion can be shifted in two dimensions to the pseudopotential node. (b) A centered ion trajectory (blue) is compared to a trajectory of a displaced ion -20μm out of the rf node (orange). The stability parameters are a $= 0.1$ and q $= 0.5$.

2.2 Multi-layer microchip trap design

The multi-layer microfabricated Paul trap is adapted from a linear Paul trap with a three-dimensional radial cross section. In quantum information science the macroscopic linear Paul traps with a pure quadrupole potential for ion trapping are designed using a single axial segment [Dre98, Sch03] - furthermore the applicability of microfabrication techniques was shown for the first time with three-dimensional traps as scalable ion microtraps with multiple control segments [Hen06, Sti06, Des06, Sch06, Ami08].

The accurate numerical simulation of the electrial potentials and the pseudopotential for ^{40}Ca$^+$ for the two-layer microchip trap is fundamental for the calculation of the characteristic trap parameters (Fig. 2.5a). The exact static potentials on the linear trap axis are essential for the simulation of ion shuttling operations with precisely shaped axial potentials (Fig. 2.5b). Beginning with the ion position at a given potential configuration, the position and orientation of the principal axes caused by the pseudopotential of the rf quadrupole characterize the static trap properties. The dynamic attributes are defined by the strength and energy of the radial and axial motional frequencies and the trap depth. The trap geometry of the radial cross section (Fig. 2.5a) is similar to a standard four rod linear Paul mass filter [Pau55]. The aspect ratio of the storage region by 4:1 leads to a de-

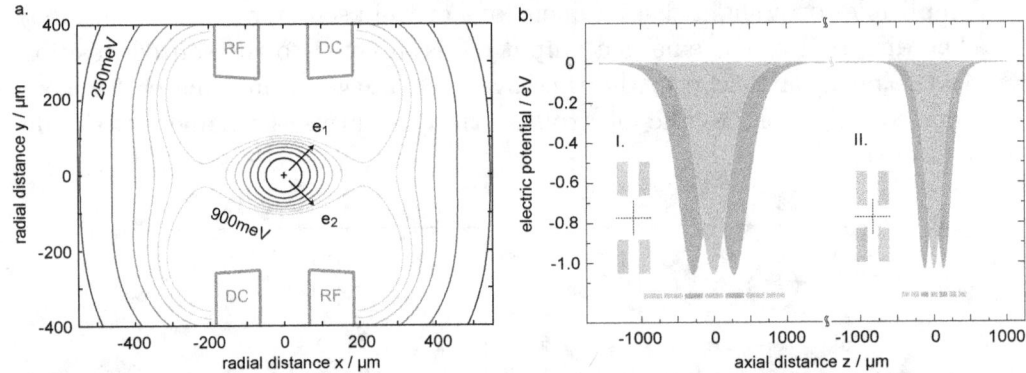

Figure 2.5: Numerical field simulations for trap operation optimized in the loading region: (a) The pseudopotential cross-section of the loading region shows equipseudopotential lines from 0.125eV to 0.75eV with a regular increment of 125meV for a $^{40}Ca^+$ ion trapped at $\Omega = (2\pi)\,24.8$MHz at $U_{rf} = 300$V. The additional pseudopotential line at 0.9eV shows the main loss channel perpendicular to the electrode layers, the trap depth in this configuration is 0.875eV. (b) The static axial potential along the trap axis causes the static confinement in the loading (I) and the processing (II) region. The independent electric potentials of three adjacent electrode pairs are superimposed at a voltage configuration of -5V (other electrodes at 0V).

formed pseudopotential with different secular frequencies for the principal axes e_1 and e_2. At the narrowed adjacent processing zone with a ratio of 2:1 both secular frequencies move nearer to degeneracy. The design supports the optical access and loading of the trap by a thermal beam, but weakens the effective trap depth in the loading zone. Because of its asymmetric geometry the microfabricated ion trap is more suitable for operations on linear ion chains confined at the linear trap axis than experiments with large three-dimensional ion crystals.

The geometric factor α determines the quadrupole potential strength and is calculated to $\alpha = 0.52 \cdot 10^7 \text{m}^{-2}$. The processing region shows a stronger confinement with $\alpha = 1.99 \cdot 10^7 \text{m}^{-2}$. Under typical operating conditions, the dimensionless stability parameter $q = 2eU_{rf}/m\,\Omega^2$ results in $q = 0.28$ for the storage region. An averaged secular frequency $\omega_{rad} = (2\pi)\,2.52$MHz is expected. The axial potential along the trap axis is calculated in a numerical three-dmensional electric potential simulation (Fig. 2.5b). The requirements for a fast ion transport are deep axial potentials with moderate control voltages and a large spatial overlap of the axial potentials from adjacent electrode pairs. The peak widths at half-height of the axial potentials are 500μm at the storage region (250μm segment width) and 264μm at the loading region (100μm segment width). For the wider segments of 250μm in the storage zone, the trap allows a tight confinement with an axial frequency of 1.20MHz at 5V only.

2.3 Planar microchip trap design

The surface-electrode trap design is fundamentally different from three-dimensional linear Paul traps. The planar trap electrodes are located on a flat surface, providing an oscillating quadrupole potential above the surface in free space - but shallower and more anharmonic compared to the multi-layer designs. The design supports the facile fabrication using standard single-layer lithography techniques. The functional principle is investigated using analytic expressions [Chi05, Rei06b, Hou08, Wes08] for a basic understanding. Several surface-electrode traps with simple linear geometries are tested experimentally [Sei06, Bri06, Bro07, Lei07, Lab08a] - the planar trap operation is limited to a single axial potential so far. Experiments using scalable linear surface-electrode ion traps with multliple adjacent axial zones for quantum information science are not realized by now.

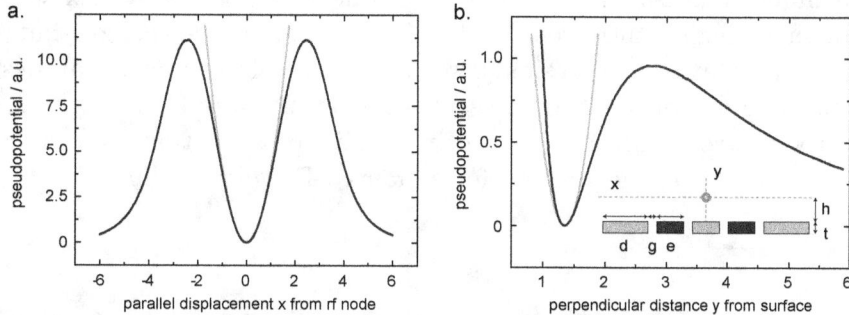

Figure 2.6: Pseudopotential cross sections of a surface-electrode trap model: The five-electrode symmetric trap design (inset) is shown at the radial cross section with the middle static electrode (orange), the rf electrodes (blue) and the outer control electrodes (orange). The geometry is characterized by e = 20 a.u., with d/e = 3, e/g = 1 and a high aspect ratio of g/t = 10. The numerically simulated pseudopotential minimum (red) is located at x = 0, y = h = 13.5 a.u. The dc electrodes (orange) are grounded in this calculation. The pseudopotential (blue) is shown in x- (a) and y-direction (b) with a parabola to illustrate the deviation from a pure harmonic radial confinement.

The deviation of the electric potentials of a surface-electrode trap compared to a three-dimensional Paul mass filter setup is obvious (Fig. 2.6). The pseudopotential at the cross section is plotted along the principal axes of motion of the symmetric five-electrode design. The dynamical confinement parallel to the surface in x-direction is highly symmetric (Fig. 2.6a). The asymmetric shape of the pseudopotential perpendicular to the surface in y-direction (Fig. 2.6b) illustrates the weak confinement and the primary loss channel. The trap depth is determined by the electrode geometry, mainly affected by the aspect ratio d/e and the distance of the rf electrodes. In a further numerical optimization the pseudopotential maximum in y-direction may be enhanced to increase the trap depth. The confinement including the rf node perpendicular to the trap electrodes is about one order of magnitude smaller than parallel to the surface. The highly assymmetric potential leads to a splitting of the secular frequency in two components [Sei06].

Using a symmetric trap geometry the Doppler cooling of the trapped ions is limited technically almost to a single principal axis directed parallel to the surface. A cooling of the perpendicular motional component is impossible because of the Doppler cooling laser parallel to the trap electrodes. A slight assymmetric rf electrode geometry effects a tilt of the principal axes to allow Doppler cooling in both directions [Chi05, Sei06]. The modified pseudopotential shows a displacement of the rf node in x-direction with a lowered trap depth and a torsion of the electric potentials caused by the electrode assymmetry. This leads to a tilt of the principal motional axes parallel to the trap surface, but the functional principle of a planar trap is remaining. The assymmetric four-electrode design is realized experimentally with a single control electrode at a dedicated side [Sei06], but in a non-segmented axial trap design. Prospectively scalable linear trap designs with multiple independent axial segments (Fig. 2.7) are favoured by a pair of control electrodes at both sides of the radial cross section, even allowing asymmetric electrode geometries for enhanced Doppler cooling.

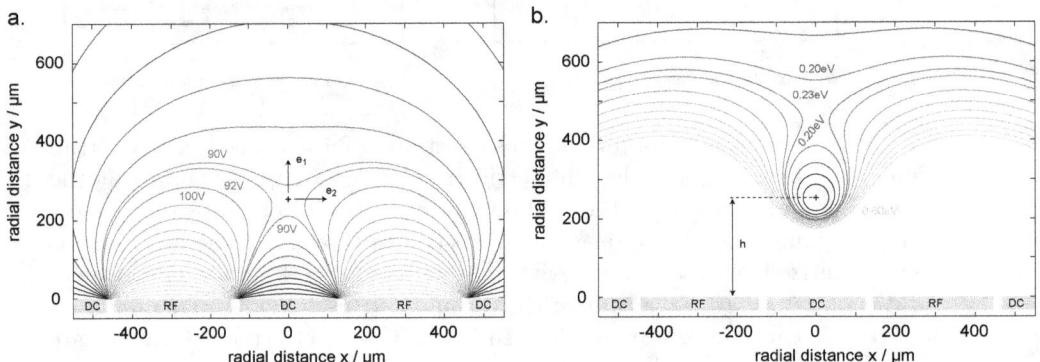

Figure 2.7: Numerical radial field simulations of the fabricated linear planar trap: (a) Transverse cross-section of the quadrupole potential corresponding to $U_{rf} = 250V_{pp}$. The trap center (indicated by the cross) with the principal axes of motion e_1 and e_2 is shown. Equi-potential lines from 10V to 240V with a regular increment of 10V and an additional line at 92V illustrate the quadrupole-shaped potential. (b) Corresponding pseudopotential to (a) for a $^{40}Ca^+$ ion trapped at $\Omega = (2\pi) \, 21MHz$. The trap depth is 0.225eV and the trap center is located at $h = 246\mu m$ above the trap surface. Equipseudopotential lines from 0.15eV to 0.6eV with a regular increment of 50meV and an additional line at 0.23eV show the main loss channel respective the weak trap confinement perpendicular to the surface.

2.4 Trap design optimization

The numerical optimization of the trap electrode geometry is illustrated on the example of the two-dimensional microchip trap (Fig. 2.8a). The electric trapping potentials are calculated separately at the two-dimensional radial cross section (Fig. 2.8b) and the one-dimensional linear trap axis based on the trap electrode geometry (Fig. 2.8c). Ideally, a pure radial dynamical quadrupole potential is reproduced approximately by the geometry of the trap electrodes as well as a static harmonic axial confinement parallel to the linear trap axis. Optimizing the electrode design of the ion trap, higher-order anharmonic terms of the trapping potential can be eliminated completely to avoid additional motional sideband excitation and power loss. The design objectives contain the optimal dimensions and aspect ratios of the trap structure and the optimal electrical trap parameters.

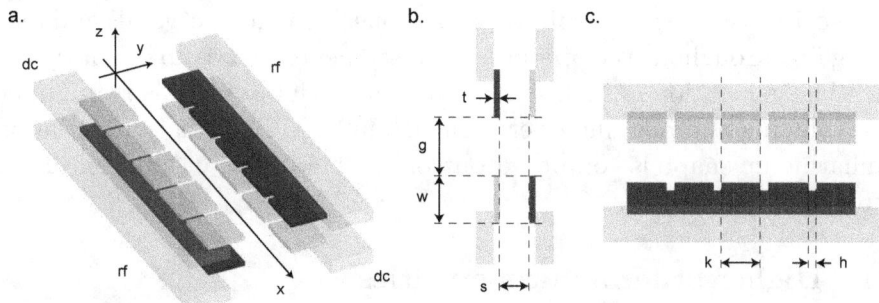

Figure 2.8: Geometric model of the two-layer microchip trap: (a) The segmented dc electrodes (gray) are located at a different layer together with a single unsegmented rf electrode (black). The rf voltage is applied at the two rf electrodes for the radial confinement, the static voltages for the axial confinement are chosen for each dc electrode separately. (b) The two electrode layers have a thickness t and are spaced by the distance s. The length of the trapping electrodes is w, the rf electrode and the segmented dc electrode are separated on each layer by the gap g. (c) The dc electrode segments are of the length k and separated by the gap h. The linear trap symmetry axis is later denoted as the x- or axial direction.

The radial trap geometry (Fig. 2.8b) is designed for a high secular frequency ω_{rad} to achieve a tight dynamical confinement of the trapped ions within the Lamb-Dicke regime. The radial confinement is characterized typically by frequencies of several MHz and should be achieved with moderate voltages on the trap electrodes of several hundred volts. The breakdown voltage caused by the fabrication technology and the width of the electrode gaps limits the operation range for the rf trap drive [Chi05, Sti06]. Additionaly the anharmonicity of the radial trapping potential is avoided by numercial optimizations. From the fact that linear traps with optimized electrode shapes have been shown to load large crystals of ions [Mor06], the loading rate and storage lifetime is improved by reducing non-harmonic contributions to the potential. Especially for a large stability parameter q at high trap drive voltages, non-linear resonances enforcing ion loss have been observed [Alh95, Alh96]. This confirms that even small anharmonicities are relevant in the case of large crystals.

In order to maintain the linear appearance of the ion crystals (Fig. 2.8c), the axial trap frequencies have to be lower than the radial frequency. Nevertheless, the axial frequencies should exceed a few MHz. Then cooling techniques are simpler [Esc03], gate operations may be driven faster and a faster adiabatic transport of ions may be realized [Sch06]. Ion transport between axial segments requires a fast update rate of the trap control voltages on the order of several μs. The fast voltage control by commercial high-speed digital-to-analog converters limits the voltage range to approximately ± 15V, which is considered at the scaling of the trap dimensions. Furthermore, scalable algorithms requires to split off single ions from linear crystals and merge them again throughout the operation of a segmented ion trap quantum computer. The generation of highly non-harmonic axial potentials is fundamental for splitting and merging operations [Hom06a]. The axial potential with an anharmonic shape is composed out of the potentials of adjacent control electrodes.

2.4.1 Higher-order anharmonicities

The quadrupole approximation of the radial potential is inaccurate if the electrode shape deviates strongly from the ideal hyperbolic form. As a result, anharmonicities and coupling terms appear inside the stability region [Alh95, Alh96]. As the radio frequency voltage is portioned to various higher-order terms and not only to the quadrupole contribution of the potential, a loss of the trap stiffness $\alpha = \alpha_2$ is observed. For simplicity the static potential U_{dc} is set to zero - the anharmonicity of the pseudopotential at the radial cross section is characterized along the two principal axes of motion. The leading terms of the polynomial expansion are specified by the radial coordinate $r^2(y, z) = y^2 + z^2$ of the cross section,

$$\phi(r(y, z), t) \propto \sum_k \alpha_k \cdot r^k \,.$$

The odd-numbered terms of anharmonicity $\alpha_1, \alpha_3, \ldots$ are excluded here, and the potential offset α_0 is ignored. The optimization of the radial trap potential leads to a suppression of the higher-order potential contribution, such that the leading non-harmonic contribution α_4 is minimized. Based on the radial optimized geometry the static axial potential at the linear symmetry axis x is analogously expanded,

$$\phi_{ax}(x,t) \propto \sum_k \beta_k \cdot x^k \ .$$

The shape of the axial potential is determined by the segmented electrode geometry, especially the axial width of the control electrodes. An optimal axial confinement of the ion requires a maximum quadratic term $\beta = \beta_2$. The single ion transport is facilitated by a large potential overlap of adjacent segments - for the splitting operation of a two-ion crystal ending in spatial separated axial potentials, a electric potential with a maximum quartic term β_4 and minor quadratic contribution β_2 is suggested [Hom06a].

The relevant parameters of various three-dimensional linear ion traps (Tab. 2.1) illustrate the frequency range, trap depth and stability parameters based on the electrode geometry. The Aarhus hexapole design with endcaps [Dre98] and the Innsbruck blade design [Sch03] show a traditional macroscopic approach of a mm-sized linear trap design without segmentation of the control electrodes. The Michigan trap designs, the microstructured three-layer trap [Hen06] and the semiconductor two-layer trap [Sti06], represent the progress in the miniaturization of linear ion traps. The numerical simulation and optimization of the radial and axial trapping potentials cover the parameter range for microstructured ion traps:

	[Dre98]	[Sch03]	[Hen06]	[Sti06]	simul.
R $[\mu m]$	1750	800	100	30	89
α $[m^{-2}]$	$1.6 \cdot 10^5$	$3.9 \cdot 10^6$	$2.2 \cdot 10^7$	$4.7 \cdot 10^8$	$5.3 \cdot 10^7$
$\Omega/2\pi$ [MHz]	4.2	23.5	48.0	15.9	50.0
U_{rf} [V]	$2 \cdot (50 \ldots 150)$	700		8	120
q	$0.2 \ldots 0.6$	0.6	0.3	0.6	0.3
$\omega_{rad}/2\pi$ [MHz]	$0.3 \ldots 0.8$	5.0	5.0	4.3	5.0
$\omega_{ax}/2\pi$ [MHz]	≤ 0.4	1.0	2.5	1.0	2.5
Δ [meV]	$\leq 10^5$	1000		80	300
	^{24}Mg$^+$	^{40}Ca$^+$	^{112}Cd$^+$	^{112}Cd$^+$	^{40}Ca$^+$

Table 2.1: Trap design comparison of several linear ion traps: The geometric trap size R given by the minimal distance between the ions position and the electrode surface and the quadratic geometry factor α of the radial cross section describes the Paul trap characteristics based on the trap geometry. The trap drive frequency Ω and voltage U_{rf} determines the q-parameter and the frequency ω_{rad} of the secular motion. The axial frequency ω_{ax} illustrates the static confinement parallel to the linear trap axis, the overall trap depth Δ summarizes the confinement of a single ion. The results of the trap simulation are compared with existing linear Paul traps used in quantum optics experiments with different ion species.

2.4.2 Radial optimization

At first the pure radial confinement of the linear trap is optimized (Fig. 2.8b). The width g of the slit is varied and the two-dimensional electric potential is calculated numerically. The electrode distance is fixed to a thickness of a standard commercial alumina wafer of $t = 125\mu$m which is used as a spacer between the two electrode layers. The variable parameter for the optimization is the slit width of the laser cut at the trap chips, respectively the distance between the rf and dc electrodes at the same layer. The radial confinement increases with a narrower slit. Interestingly, the radial potential for the two-layer electrode design is almost harmonic since the fourth order parameter α_4 is nearly vanishing for $g > 110\mu$m. For a slit width of $g = 126\mu$m the secular frequency of $\omega_{\mathrm{rad}} = (2\pi)\,5$MHz is obtained for ^{40}Ca$^+$ at a trap drive voltage of 240V$_{\mathrm{pp}}$ and a drive frequency of $\Omega = (2\pi)\,50$MHz.

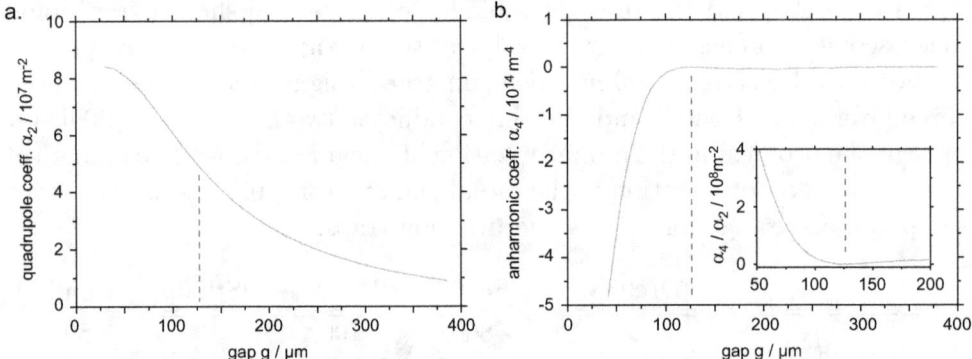

Figure 2.9: Numerical simulation of the radial trap stiffness and anharmonicity of the trap potential as function of the slit width g: (a) The quadrupole coefficient α_2 decreases with increasing slit width. (b) Starting at a small gap, the contribution of the anharmonicity parameter α_4 decreases with bigger gap size. Simultaneously the trap stiffness α_2 is lowered. The ratio between the first anharmonicity coefficient α_4 and α_2 shows the harmonic optimum at a slit width of $g = 126\mu$m (dashed line), corresponding to a quadrupole parameter of $5.5 \cdot 10^7$m^{-2}.

The result (Fig. 2.9) of the optimization illustrates the empirical constraint of $g/t \geq 4/5$ for a non-significant contribution of the anharmonicity α_4. The non-linear progression of the trap stiffness α_2 to a smaller gap size g regarding this constraint covers a broad range of technical realizable aspect ratios. Comparing the electrode geometry of the fabricated microchip trap with the numerical simulations, the slit is decreased from 500μm to 250μm from the storage to the processing zone, the trap stiffness may be increased by one order of magnitude ending with a slit size of 126μm. Then technical limitations like laser background scattering for detection and a dynamical q-adaption by varying the trap drive voltage have to be considered.

2.4.3 Axial optimization

The optimization of the axial potential determines the performance of ion shuttling operations. Additional requirements are the generation of deep axial potentials even with moderate control voltages and the capability of the splitting of a linear two-ion crystal ending with spatial separated ions located at distinct potential minima.

At first the axial trap frequency ω_{ax} is maximized by the variation of the segment width k at a constant spacing of h $= 30\mu$m (Fig. 2.10). The three-dimensional numerical calculated electric potential is more shallow for larger control electrodes, and for a very short width k the electric potential falls rapidly off from the electrode tips resulting in a weak confinement. The maximum axial trap frequency is achieved for a segment width of k $= 70\mu$m, the radial cross section is characterized by the optimized gap size of g $= 126\mu$m. Changing the size of the segment width k by 50% results in a 20% variation of β_2 which is easily compensated by the control voltages.

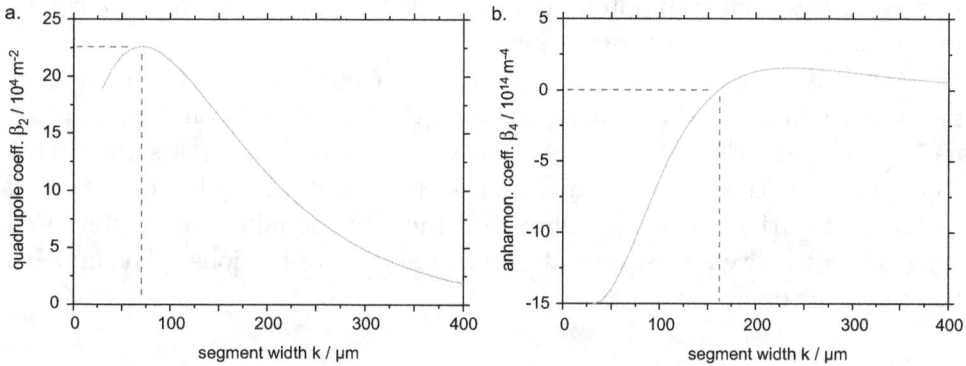

Figure 2.10: Numerical simulation of the axial trap stiffness and anharmonicity of the trap potential: (a) At the maximum of the coefficient β_2 the trap potential is anharmonic. By increasing the segment width k, the anharmonicity coefficient β_4 converges to zero (b), but the quadrupole coefficient is reduced strongly. At a segment width near 160μm the trap potential is nearly harmonic because of the vanishing anharmonicity parameter β_4.

The storage and processing of single ions requires a different electrode configuration compared to the realization of efficient transport operations. For ion shuttling, a large overlap of the axial electric potentials is important. The storage operation is obtained by biasing two adjacent electrode pairs with an equal voltage to obtain a virtually a larger electrode. Due to the smaller segmentation an enhanced overlap of the individual potentials is provided for the transport operations. An optimal effective segment width is determined by minimizing the anharmonicity term β_4. The fabricated microchip trap is realized with a segment width of k $= 250\mu$m in the storage and 100μm in the processing region.

2.5 Ion shuttling operations

Detailed simuations for large-scaled quantum processors show that up to 99% of the operating time for quantum algorithms will be spent with transport processes [Chu]. The time required for the transport should be reduced to improve the gate times and reduce the decoherence processes. In recent experiments [Row02, Lei05, Hub08] the shuttling has been carried out within the adiabatic limit, such that the time required for the transport by far exceeds the oscillation time of the ion in the axial potential. It is a common misbelief that this adiabatic transport is necessary to avoid the excitation of vibrational quanta. The fast and non-adiabatic single ion shuttling is investigated by applying classical optimal control theory.

Starting with realistic numerical simulated axial potentials, the optimized transport is characterized by none excitation of oscillating motional degrees of freedom, even though the transport speed exceeds the adiabatic regime. A single ion transport is achieved within roughly two oscillation periods in the axial trap potential, compared to typical adiabatic transports including an order of 10^2 oscillations.

Certainly, non-optimized fast transport of qubit ions followed by sympathetic cooling of a different ion species [Bar03, Dre04] would be an alternative strategy. However, the necessary cooling time increases the overall computational time. First experiments show that the qubit coherence is maintained during a transport, but that the vibrational quantum state may typically not be well-conserved after a fast shuttle of the ions. This impedes further qubit operations.

2.5.1 Non-adiabatic heating sources

The transport is calculated starting at the center of a control electrode pair. The trap design is adapted from the two-layer microchip trap design (Fig. 2.8). The single ion shuttling ends at the center of the next electrode pair - the simulation can be scaled up adequately to larger shuttling protocols. The goal of the optimization is a decrease of the transport time below the limit of adiabaticity, such that the transport is finished within a single oscillation period only - with the constraint to avoid vibrational excitation. The micromotion on the linear trap axis is neglected, the ion is moved in the radial rf node. The potential minimum caused by the control electrodes is shifted by changing the dc voltages $u_i(t)$ (Fig. 2.11). Intuitively, a smooth acceleration and deceleration for the ion is advantageous. The non-adiabatic heating due to the fast transport has to be minimized.

The non-adiabatic heating sources appears for transports on a timescale of the axial trap frequency and are characterized as a classical displacement error, wavepacket dispersion heating and parametric heating. The classical displacement error describes the deviation between the ions position and the potential minimum - in the classical point of view the ion starts oscillating and at the end of the transport excess energy is remaining. In the quantum picture it corresponds to the buildup of a non-vanishing displacement during the transport. The wavepacket dispersion heating is originated by the anharmonicity of the axial potential, which results in vibrational excitation. The effect is nonsignificant here, because the spatial extension of the wavepacket is about 10nm to 20nm and the undisplaced wavepacket senses hardly potential anharmonicities generated by 50μm sized electrodes. The source of parametric heating is significant, if the wavepacket width can't follow the variation of the axial frequency ω_{ax} adiabatically. Then parametric heating to higher vibrational states will occur.

2.5.2 Non-adiabatic transport optimization

The optimization algorithm deals with the minimization of the classical displacement error by applying optimal control techniques [Kir04]. The single ions axial trajectory is optimized regarding to the entirely classical error source, such that the cost function - weighting the phase space displacement after the transport - is minimized (Fig. 2.11). The parametric heating is suppressed by an appropriate initial guess which keeps the trap frequency perfectly constant. This is achieved by a variable transformation from $u_{1,2}(t)$ to new parameters that allow to decouple the strength of the potential and its minimum position (Fig. 2.13). Starting now the optimal control method yields a solution that reduces the displacement error. Since the control parameters are modified slightly by the optimization algorithm [Cal04, Dor05], the parametric heating and also the wavepacket dispersion heating are negligible.

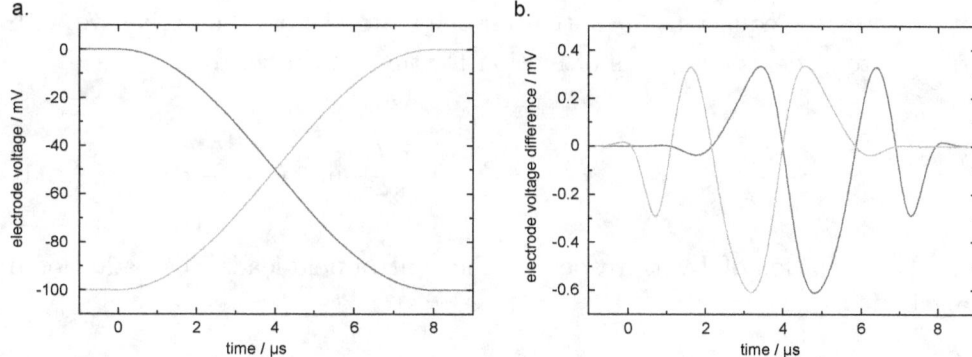

Figure 2.11: Initial guess (a) and residual optimization (b) for the transport between two electrode pairs: The single ion transport begin at the start electrode (orange) at 0μs and stops at the target electrode (gray) at 8μs. The residual optimized control voltages are given with respect to the time-dependent initial guess voltages.

The calculation method is derived from a variational principle with unbounded controls and fixed final time [Kir04]. The axial motional state of the single ion is confined in the ground state[2] and is represented by $\tilde{\xi}(t) = (x, v)$ in a two-dimensional phase space. The equation of motion follows to

$$\dot{\tilde{\xi}} = \tilde{a}(\tilde{\xi}, \{u_i\}) = \begin{pmatrix} v \\ -\frac{1}{m} \sum_i \frac{\partial V_i(x)}{\partial x} u_i(t) \end{pmatrix} , \tag{2.9}$$

with $i = 1, 2$ for both electrode pairs and $V_i(x)$ as the normalized electrode potentials. The goal is to find the time-dependent control voltages $u_i(t)$ to move the single ion from the center of first electrode to the second electrode center. We desire to have the ion at rest after the transport process. The performance of a given control field is judged by the cost function

$$h(\tilde{\xi}(t_f)) = \alpha \left(x(t_f) - x_f \right)^2 + \beta\, v(t_f)^2 , \tag{2.10}$$

which measures the phase space displacement at the final time t_f. The constants α and β weight the contributions to each other. The equation of motion (2.9) is a constraint for all times t, so the following cost functional is obtained by

$$J(\tilde{\xi}, \tilde{\xi}_p, \{u_i\}) = \int_0^{t_f} \frac{\partial h}{\partial \tilde{\xi}} \cdot \dot{\tilde{\xi}} + \tilde{\xi}_p \cdot \left(\tilde{a}(\tilde{\xi}, \{u_i\}) - \dot{\tilde{\xi}} \right) dt$$

with the costate vector $\xi_p = (x_p, v_p)$ as a Lagrange multiplier in order to guarantee that the optimization result obeys the equation of motion. The time dependence of all variables has been dropped in the notation. For an optimal control field, $\delta J = 0$ has to hold, therefore the variational derivatives with respect to $\tilde{\xi}, \tilde{\xi}_p$ and $\tilde{u}$ have to vanish. The derivative with respect to $\tilde{\xi}_p$ restores the equations of motion for the state vector, the derivative with respect to $\vec{\tilde{\xi}}$ yields equations of motion for the costate vector:

$$\dot{\tilde{\xi}}_p = -\frac{\partial \tilde{a}}{\partial \tilde{\xi}} \cdot \tilde{\xi}_p \;\Rightarrow\; \dot{x}_p = v_p \frac{1}{m} \sum_i \frac{\partial^2 V_i(x)}{\partial x^2} u_i \;,\; \dot{v}_p = -x_p \tag{2.11}$$

The variation of J with respect to the control field leads to an additional algebraic equation:

$$\frac{\partial \tilde{a}}{\partial u_i} \cdot \tilde{\xi}_p = 0 \;\Rightarrow\; -\frac{1}{m} \frac{\partial V_i(x)}{\partial x} v_p = 0. \tag{2.12}$$

The boundary condition $|\partial h/\partial \tilde{\xi}|_{t_f} = 0$ for $\tilde{\xi}_p$ is derived by variation with respect to the final state. If the ion starts centered in the potential well

---

[2]The calculation is valid also for thermal and coherent states with modest excitation.

of the first electrode pair, the set of boundary conditions for the state and costate vector reads

$$x(0) = 0 , \; v(0) = 0 , \; x_p(t_f) = 2 \, (x - x_f) , \; v_p(t_f) = 2 \, v . \qquad (2.13)$$

The equations (2.9), (2.11) and (2.13) together with (2.12) represent a system of coupled ordinary nonlinear differential equations with splitted boundary conditions, i.e. for two of the variables, initial conditions are given whereas for the other two, the values at the final time are specified. This makes a straightforward numerical integration impossible. The system is therefore solved in an iterative manner by means of a gradient search method. The scheme of this steepest descent algorithm is as follows:

1. Choose an initial guess for the control field $u_i(t)$.

2. Propagate x and v from $t = 0$ to $t = t_f$ while using $u_i(t)$
   in the equations of motion. At each time step $x(t)$ is saved.

3. Determine $x_p(t_f)$ and $v_p(t_f)$ according to (2.13).

4. Propagate x_p and v_p backwards in time from $t = t_f$ to $t = 0$.
   At each time step, save the value of $v_p(t)$.

5. For each time step, the control field with the voltage V_1
   at the first electrode is updated according to

$$u_i^{new}(t) = u_i^{old}(t) + \tau \, v_p \, \frac{1}{m} \frac{\partial V_1(x)}{\partial x} \qquad (2.14)$$

6. Repeat steps 2 to 5 until the specified fidelity is reached.

The gradient search step width τ and the parameters α and β are chosen empirically in equation (2.14). If it is too small, the algorithm converges too slowly, if it is too large, the algorithm starts to oscillate. For the presented results the following values $\alpha = 10$, $\beta = 1$ and $\tau = 5 \cdot 10^{-8}$ are chosen. The algorithm converged at about 200 iterations. The initial guess of the control field provides a smooth and symmetric acceleration and deceleration of the ion (Fig. 2.11a), the potential minimum coincides exactly with the desired positions at the initial and final time:

$$u_0^{(0)}(t) \;=\; \begin{cases} V_0 & \text{for } t \leq 0 \\ V_0 \, \sin^2(\frac{\pi t}{2\Delta t}) & \text{for } 0 < t \leq \Delta t \\ 0 & \text{for } t > \Delta t \end{cases} \qquad (2.15)$$

$$u_1^{(1)}(t) \;=\; V_0 - u_0^{(0)}(t)$$

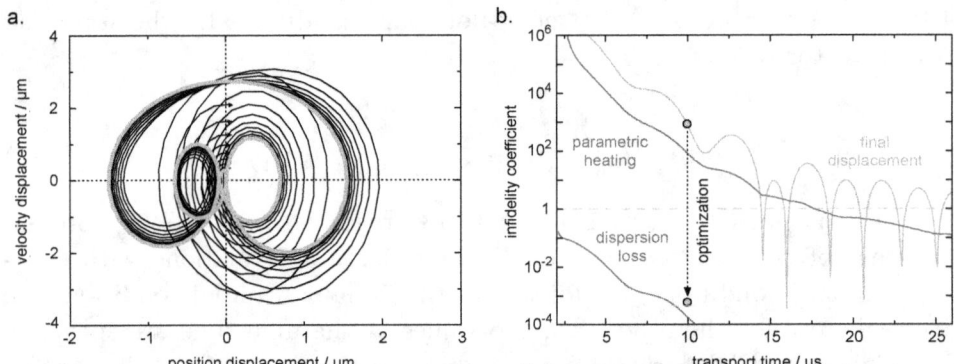

Figure 2.12: Phase space inspection and optimization results: (a) The phase space trajectories in the frame of the co-moving with the potential minimum are shown for the iteration $0, \ldots, 100$. The ion arrives close to the phase space origin. (b) Indicated is the excess energy as a function of the transport time. The final displacement describes the energy in vibrational quanta. The heating of a single phonon (dashed gray line) permits the discrimination to non-adiabatic behavior referring to the parametric heating. The phase space displacement for the initial guess and the improvement for the optimized control voltages are shown (circles).

In principle, other initial guess voltages like Gaussian waveforms can be used as well. The reference voltage is $V_0 = -0.1V$, corresponding to $\omega_{ax} \approx (2\pi)\,0.5\mathrm{MHz}$ at the initial and final potential. The discrete time step is sized to $\Delta t = 8\mu s$, the total shuttling time lasts from $-1\mu s$ to $9\mu s$. Regarding to the ion heating due to anharmonic dispersion, the system is described quantum mechanically by a Hamiltonian of a harmonic oscillator with an anharmonic dispersion:

$$\mathrm{H}_0(t) = \frac{\hat{p}^2}{2m} + \frac{m\,\omega_{ax}(t)^2}{2}(\hat{x} - x_0(t))^2 + \kappa(t)(\hat{x} - x_0)^4$$

Without the temporal variation of ω_{ax} and the anharmonicity $\kappa(t)$ of the potential, the solution of the time-dependent Schrödinger equation is a coherent state $|\alpha(t)\rangle$, where the displacement parameter $\alpha(t)$ is inferred from the classical trajectory. Anharmonic dispersion of a wavepacket occurs at a timescale given by $\mathrm{T}_{rev}/(\Delta n)^2$ [Tan06], with the revival time

$$\mathrm{T}_{rev} = 2h\left(\frac{d^2E_n}{dn^2}\right)^{-1} \tag{2.16}$$

and the spread over the vibrational levels $\Delta n = \alpha(t)$. The shift of the energy levels E_n induced by the anharmonic contribution causes a finite dispersion time and can be calculated in first order stationary perturbation theory:

$$\Delta E_n(t) = \frac{5}{4} \frac{\hbar^2 \kappa(t)}{m^2 \omega_{ax}(t)^2} n^2 .$$

A generalized dispersion parameter can be defined - if this parameter is sufficiently small, anharmonic dispersion will not contribute to heating:

$$\int_0^{t_f} dt \, \frac{\Delta n^2}{T_{rev}} = \frac{5\hbar}{4\pi m^2} \int_0^{t_f} dt \, \frac{\kappa(t) |\alpha(t)|^2}{\omega_{ax}(t)^2} .$$

In a quantum mechanical estimate of non-adiabatic heating the evolution of the wavepacket width is investigated, whether it follows adiabatically with the axial harmonic frequency $\omega_{ax}(t)$. If the adiabatic theorem is fulfilled,

$$\hbar \langle \phi_m(t) | \dot{\phi}_n(t) \rangle \ll |E_n(t) - E_m(t)| ,$$

transitions between eigenstates can be neglected. The parametric time dependence of the eigenstates states in 2.17 is the implicit time dependence via $\omega_{ax}(t)$. The following non-vanishing matrix elements are found:

$$\langle \phi_{n+1} | \phi_n \rangle = \frac{\dot{\omega}_{ax}}{\sqrt{2\pi^3} \omega_{ax}} n \sqrt{n+1} \tag{2.17}$$

$$\langle \phi_{n+2} | \phi_n \rangle = \frac{\dot{\omega}_{ax}}{4\omega_{ax}} \sqrt{(n+1)(n+2)}$$

and similar expressions for $m = n - 1, n - 2$. Thus, parametric heating can be neglected if

$$n^{3/2} \frac{\dot{\omega}_{ax}}{\omega_{ax}^2} \ll 1 . \tag{2.18}$$

Numerical evaluation of the matrix elements yields the result, that the adiabatic condition is fulfilled for $n = 0$, but is clearly violated for high n occurring for large excursions of the wavepacket, for example $\bar{n} \approx 2000$ for $\Delta x = 1\mu m$ at a transport time of $10\mu s$ (Fig. 2.12).

Therefore the optimization strategy has to be refined to keep the axial frequency $\omega_{ax}(t)$ constant (Fig. 2.13). The control voltages are symmetric, so one control degree of freedom can be sacrificed. Then the initial guess voltages (2.16) are normalized to a constant $\dot{\omega}_{ax} = 0$ before the optimization. The optimization process leads to variations in $\omega_{ax}(t)$ that are negligibly small - typically leading to maximum values of $\dot{\omega}_{ax}/\omega_{ax}^2$ on the order of 10^{-5}. The adiabatic theorem is fulfilled according to (2.18) even after the optimization algorithm has cured the classical phase space displacement heating. This is in strong contrast to the unconstrained, previous guess function, where we obtain $\dot{\omega}_{ax}/\omega_{ax}^2 \simeq 10^{-2}$. It should be noted that parametric heating can be completely suppressed as well as for the optimized control voltages. This can be achieved by changing the set of control parameters to $\tilde{u}_1 = u_1 + u_2$

and $\tilde{u}_2 = u_1/\tilde{u}_1$. The parameter $\tilde{u}_2$ is directly related to the instantaneous potential minimum x_0. If $\tilde{u}_2$ is incorporated in the optimization only, $\tilde{u}_1$ can be readjusted at each step to keep ω_{ax} constant.

The optimization results for the improved initial guess voltages are shown in Fig. 2.14. The transport time could be reduced to $5\mu s$ which corresponds to roughly two oscillation periods. For the improved guess funtion the wavepacket dispersion appears now as the dominant heating source. This process could be suppressed either by further geometric optimization of the trap segments or by including a corresponding additional term into the cost function of the optimization routine. Further the unwanted heating can be suppressed by many orders of magnitude by the application of appropriate time-dependent control voltages.

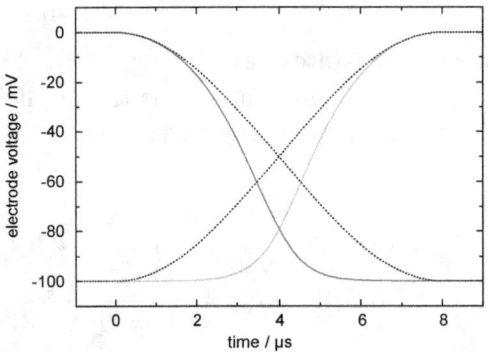

Figure 2.13: Initial guess voltages normalized to constant axial trap frequency. The start electrode (orange) and the end electrode voltage (gray) are compared to the old initial guess voltage (dashed). The dynamics of the potential minimum is unaffected by the normalization.

Technically, this could be achieved by using fast high-resolution digital-to-analog converters. The small correction voltages obtained from the optimization algorithm might represent a problem, however a 16bit converter allows a discretization of roughly $1.5\mu V$ for a maximum voltage change of 0.1V. The robustness of the solution is checked against noise by calculating the trajectories with white noise of variable level added on the voltages. A quadratic dependence of the excess displacement on the noise level is found. The deviation of the final displacement from the noise-free case was negligibly small at a noise level of $20\mu V$. Experimental values for non-adiabatic heating effects in ion transport are given in [Row02]. The comparison with our theoretical values is hampered by the fact that these measurements have carried out at higher axial trap frequency and the lighter ion species $^9\mathrm{Be}^+$, but on a much longer transport distance of 1.2mm. However, low heating rates were obtained in those experiments only if the transport duration corresponds to a relatively large number of about $\simeq 100$ trap periods, whereas in our case, a transport within only two axial periods was simulated.

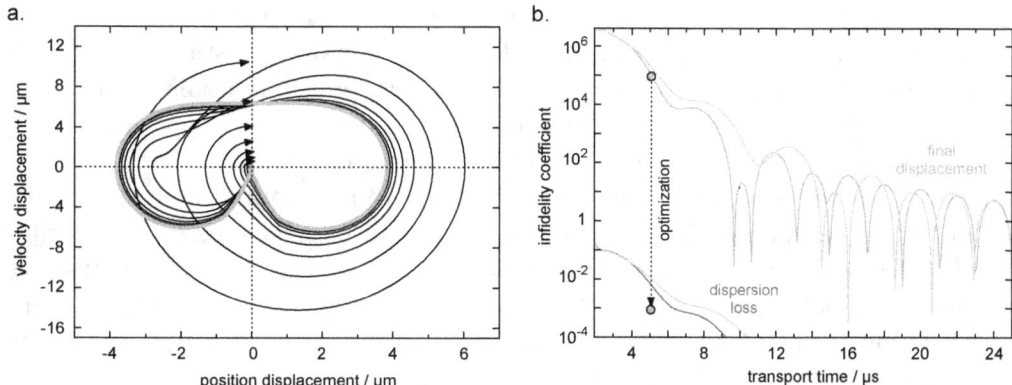

Figure 2.14: Optimization results with improved initial guess: (a) The ion arrives close to the phase space origin. The optimiziation is calculated for a transport time of $5\mu s$ - corresponding to about two axial oscillation periods in the harmonic trap potential. (b) Non-adiabatic effects versus transport time for the improved initial guess voltages. The final displacement (orange) and the dispersion parameter (gray) is shown (old initial guess is grayed), parametetric heating is not relevant anymore. The improved initial guess allows a transport time of $5\mu s$ and an optimization of about eight orders of magnitude in classical phase space displacement. Now, with a few μs transport time, anharmonic dispersion becomes the predominant heating source.

The optimization of ion transport beyond the speed limits given by the anharmonic terms of the trapping potentials and parametric heating would be most efficient if a full quantum mechanical equation of motion will be employed. Quantum mechanical optimal control methods are based on the same variational principle as presented for a classical problem, with the only difference that the terms in the penalty functional are functionals on Hilbert space. Algorithms for quantum mechanical optimal control are well developed and were applied to a variety of problems [Skl02, Cal04].

In this case however, the application of quantum mechanical optimal control was not yet possible for simply a technical reason: The iterative solution via repeated solution of the Schrödinger equation over distances on the order of $200\mu m$ and time intervals on the order of $20\mu s$ takes too much computational effort. On the other hand, it is obvious that for the typical potentials of Paul traps, the possibility to exert quantum control on the system is very restricted since the wavefunction of the ion mainly senses a harmonic potential. The classical approach is therefore well suited to the problem.

The application of quantum mechanical optimal control methods also opens new possibilities, for example the control voltages could be used to devise schemes for quantum computational gates. In this case, the target wave function for the optimization routine could be the first excited motional state or even a superposition of different motional Fock states. To fulfill this aim, anharmonic contributions to the trapping potentials are crucial. Another promising strategy that could be employed to avoid heating during ion transport is the closed loop control technique. Here, the experimental results are fed back into an evolutionary algorithm to obtain improved values of the control parameters. The heating rate can be measured by comparing the strengths of the red and blue motional sidebands after the transport process [Roo99]. The key problem for applying closed loop control to ion shuttling is the appropriate parametrization of the control voltages in order to keep the parameter space small.

This work has started to apply the optimal control theory for ion trap based quantum computing. Not only the motion of ions between trap segments, but the entire process including shaped laser pulses [Gar03] and motional quantum state engineering might be improved with this technique.

Chapter 3

$^{40}\mathrm{Ca}^+$ as the qubit candidate

For ion trap operation and characterization the ionized isotope $^{40}\mathrm{Ca}^+$ is used as the qubit candidate. The ions are generated by photoionization from an effusive thermal beam of neutral calcium. The calcium isotope $^{40}\mathrm{Ca}$ is the most naturally occurring isotope with 97% - pure Calcium is artificially because of its exceptional chemical reactivity. In an ultra high vacuum environment the solid calcium with the lowered vapor pressure is evaporated at temperatures of $> 350°\mathrm{C}$.

The optical level scheme of the $^{40}\mathrm{Ca}^+$ isotope (3.1) for quantum state readout and qubit initialization is accessible with commercial diode lasers. In contrast to many other ion species used in quantum information processing no dye laser or solid state laser, nor a high power pump laser is required, so all optical transitions are operated using diode laser systems. For the photoionization of the neutrals two diode lasers are used, the Doppler transitions of the ion are driven by three diode lasers. Two different ways for qubit preparation (3.2) of an optical or spin qubit are presented - with a tapered amplified diode laser at the quadrupole transition for an optical qubit, or optionally with a frequency-doubled tapered amplified diode laser driving Raman transitions at the Zeeman sublevels of the ground state for a spin qubit.

3.1 Level scheme and transition parameters

The atomic properties of the $^{40}Ca^+$ ions are founded by the absent nuclear spin, that leads to a nonexisting hyperfine structure like the other isotopes except $^{43}Ca^+$, which has a nuclear spin of 7/2. The electronic configuration of the ion is similar to the neutral earth alkali metals, here with a single additional electron close to the nobel gas configuration of Argon.

The characteristic properties of the earth alkali metals with the unoccupied 3^2D-levels of the lower shell among the 4^2S- and 4^2P-levels result in a hydrogen-like configuration (Fig. 3.1). The 3-level system has a Λ-configuration with the $D_{3/2}$, $D_{5/2}$ and $S_{1/2}$ levels coupled by dipole transitions to the common levels $P_{1/2}$ and $P_{3/2}$. The $D_{3/2}$ and $D_{5/2}$ levels are metastable and the $S_{1/2}$ level is the ground state, so by an excitation with equal frequency detuning of both dipole transitions to the $P_{1/2}$ level a coherent superposition of the two lower levels $S_{1/2}$ and $D_{3/2}$ can be generated. Because of the Raman-type coherent excitation the common level remains unpopulated and the fluorescence is suppressed for an equal detuning. In the fluorescence spectrum of the $S_{1/2} \leftrightarrow P_{1/2}$ transition a dip called dark resonance appears. The shape and width of this Raman resonance is determined by the linewidth and power of the lasers driving both Doppler transitions. Respecting the Zeeman sublevels of the Λ-system, the ground state $S_{1/2}$, the metastable state $D_{3/2}$ and the $P_{1/2}$ level has two, four and two components, respectively. The 12 transitions dependent by the polarization of an existing magnetic field result in 8 detectable Raman resonances because of four degenerate transitions.

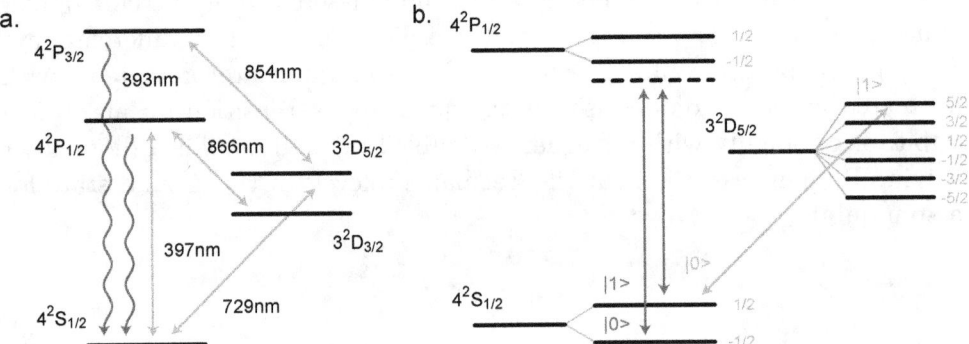

Figure 3.1: Level scheme and different qubit systems for $^{40}Ca^+$: (a) The lowest energy levels define the relevant transitions based on the electronic ground state $4^2S_{1/2}$. A single UV laser at 397nm and three lasers at 729nm, 854nm and 866nm control the level population, the state detection is located in the UV range at 393nm. (b) The Zeeman energy splitting in a magnetic field is shown depending on the magnetic quantum number m_J (gray) and the individual Landau factor g_J for each level. The qubit states of the optical and spin qubit schemes are denoted as $|0\rangle$ and $|1\rangle$.

The wavelength λ for the Doppler and the quadrupole transitions are in the UV and IR range, the metastable transitions has the longest lifetime τ of the five lowest energy levels [Jam98]:

	$S_{1/2} \leftrightarrow P_{1/2}$	$S_{1/2} \leftrightarrow P_{3/2}$	$P_{1/2} \leftrightarrow D_{3/2}$	$P_{3/2} \leftrightarrow D_{3/2}$
λ/nm	396.847	393.366	866.214	849.802
τ/ns	7.7(2)	7.4(3)	94.3	901

	$P_{3/2} \leftrightarrow D_{5/2}$	$S_{1/2} \leftrightarrow D_{5/2}$	$S_{1/2} \leftrightarrow D_{3/2}$
λ/nm	854.209	729.147	732.389
τ/ns	101	$1.045 \cdot 10^9$	$1.080 \cdot 10^9$

Table 3.1: Doppler and quadrupole transition wavelengths for ^{40}Ca$^+$

The energy splitting ΔE of the level fine structure by a magnetic field is determined by the Lange factor g_J, the magnetic field strength B, m_J given by the total angular momentum quantum number J and the Bohr magneton μ_B as the proportional constant:

$$\Delta E = g_J\, \mu_B\, B\, m_J \;,\; g_J = 1 + \frac{J(J+1) + S(S+1) - L(L+1)}{2J(J+1)}$$

The Lande factors g_J for the different atomic levels effects the various strength of the Zeeman shift for each level and are calculated to:

	$4^2S_{1/2}$	$4^2P_{1/2}$	$4^2P_{3/2}$	$3^2D_{3/2}$	$3^2D_{5/2}$
g_J	2	2/3	4/3	4/5	6/5

Table 3.2: Lande factors for different levels of ^{40}Ca$^+$

The five lowest energy levels of ^{40}Ca$^+$ (Fig. 3.1a) show the laser-driven transitions (orange) and the levels for fluorescence detection (gray). The 729nm laser drives the quadrupole transition, for the electron shelving scheme the laser at 854nm is used for level depletion to detect the fluorescence at 393nm. The 397nm laser for Doppler cooling effects the fluorescence at 397nm, the 866nm laser is used for repumping out of the specific metastable state. The level system can be interpreted as an idealized three level system (Fig. 3.1b). Two different qubit preparations are shown, the quadrupole transition (orange) is driven by the 729nm laser, the Raman transition to the virtual level initializes the Zeeman sublevels of the ground state as qubit states.

3.2 Quantum state preparation

Two different initialization schemes for qubit preparation are realized so far - for an optical and a spin qubit. The qubits are coupled by a quadrupole or Raman transition and distinguished by a long lifetime for an enhanced coherence during the application of various quantum algorithms. All experimental results regarding the trap characterization are realized using the quadrupole transition.

3.2.1 Quadrupole transition

The metastable quadrupole transition $|S_{1/2}, m_J = 1/2\rangle \leftrightarrow |D_{5/2}, m_J = 5/2\rangle$ initializes the optical qubit (Fig. 3.1b). Beside the coherent manipulation of the qubit, the transition is used for 729nm sideband cooling to the motional ground state and quantum state readout via quantum jump spectroscopy. Because of the different Lande factors g_J and magnetic quantum numbers m_J of the two Zeeman states the transition is first order sensitive on fluctuations of magnetic fields, which affect the coherence of the qubit.

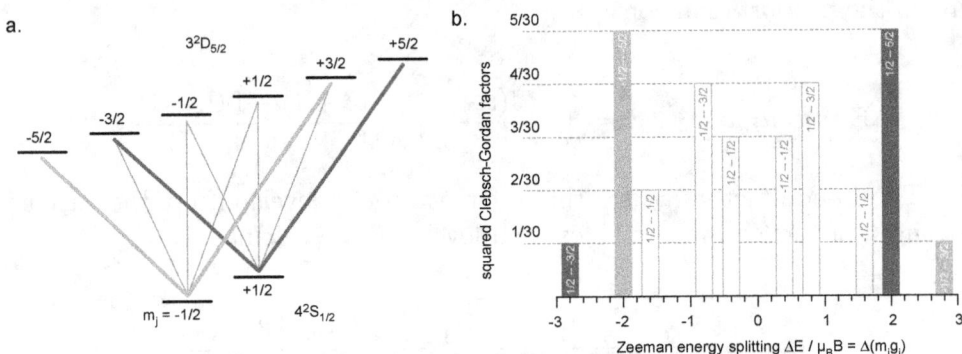

Figure 3.2: Zeeman energy splitting of the $S_{1/2} \leftrightarrow D_{5/2}$ transition in a magnetic field: (a) The quadrupole transition splits into ten different transitions with $\Delta m = 0, \pm 1, \pm 2$. The $\Delta m = \pm 2$ transitions are emphasized (orange, gray), the other transitions are not used in the experiment. (b) The maximum line strengths are proportional to their squared Clebsch-Gordan coefficients and are plotted against the Zeeman energy splitting represented by $\Delta(m_J g_J)$ in units of Bohrs magneton μ_B.

The quadrupole transition is splitted into ten components at a weak magnetic field, depending on the Zeeman sublevels of the ground state $S_{1/2}$ and the metastable level $D_{5/2}$ [Jam98]. The angle θ between the direction k of the 729nm laser and the magnetic field direction, the angle γ between the laser polarization p and the magnetic field B determine the geometrical allowed transitions. The configuration $\theta = 45°$, $\gamma = 0°$ is prerequisite for $\Delta m = 0$ transitions, the geometry based on $\theta = 90°$, $\gamma = 0°$ allows $\Delta m = \pm 1$

and with $\theta = 90°$, $\gamma = 90°$ the transitions $\Delta m = \pm 2$ are achieved.

3.2.2 Raman transition

A different approach to establish a qubit is the coupling of the Zeeman sublevels of the ground state $S_{1/2}$ by a coherent Raman transition below the $P_{1/2}$ level (Fig. 3.1b). The Zeeman sublevel energy splitting is determined by the ambient magnetic field. The coherence of the spin qubit $|0\rangle = |S_{1/2}, m_J = -1/2\rangle$ and $|1\rangle = |S_{1/2}, m_J = 1/2\rangle$ is independent from first order magnetic field fluctuations. Beside driving Raman transitions between the Zeeman sublevels, Raman sideband cooling to reach the motional ground state can be established, while the state detection can be realized by quantum jump spectroscopy using the 729nm laser with the subsequent shelving of the metastable level.

Similar to the qubit initialization with the 729nm laser, the Raman laser setup defining the geometry, power and detuning to the $P_{1/2}$ state is crucial for the Raman transition rate, the light shifts and the spontaneous emission rate due to non-resonant excitations [Sac00, Win03, Oze07, Ben08b]. The coherent coupling between the Zeeman sublevels of the ground state is realized by stimulated Raman transitions. The two Raman laser beams are detuned from the optical resonance and are perpendicular to each other, with a resulting k-vector parallel to the linear trap axis. For single qubit operation a third copropagating Raman laser is needed. The small energy separation between the two states of typically 20MHz is suitable for long-time phase coherence. The Raman detuning Δ to the $S_{1/2} \leftrightarrow P_{1/2}$ transition is several tens of GHz. The Raman detuning Δ can be positive for a blue detuning or negative for a red detuning. The Rabi frequency of the Raman setup follows to $\Omega_1 \Omega_2 / \Delta$ with the single Rabi frequencies Ω_1 and Ω_2 of the individual Raman beams. Because of $\Delta \gg 20$MHz, the detuning Δ is approximately constant for the Zeeman transition, so an effective two-level system is initialized.

Chapter 4

Atom-light interactions

The interaction of atoms with light is the exclusive mechanism for the quantum state preparation, manipulation and detection of a single trapped ion in this experimental setup. The Doppler cooling of the ions to the Lamb-Dicke regime and subsequent sideband cooling on the quadrupole transition to the motional ground state of the trapped ion are discussed. This preparation to a pure motional quantum state in the strong binding regime of the harmonic confinement is treated quantum mechanically, illustrating the coupling strength of different motional sidebands that are resolved by spectroscopic measurements using quantum jump spectroscopy.

The difference of a free and trapped atom based on a electronic two-level system is discussed (4.1), showing the interaction of the atom with the light field equivalent to a Jaynes-Cummings description. Both cooling techniques (4.2), Doppler cooling and resolved sideband cooling are fundamental for preparing the ions quantum state.

4.1   Free and trapped two-level atoms

The fundamental concepts for laser cooling regarding to the preparation of the ions quantum state adapt the techniques for free particles on trapped single ions in free space using electrodynamic fields. Starting with a thermal beam of free neutral atoms, the trapped ions are produced using photoionization and then are Doppler cooled from temperatures of hundreds of Kelvin. Subsequently, the translationally cold trapped ions are transferred to the motional ground state using sideband cooling.

Initially Doppler cooling with light was proposed for free atoms [Hae75] and electromagnetically trapped ions [Win75], the first experimental realizations were achieved with ions in Paul traps [Neu78, Win78, Win79, Ita82, Neu80]. Based on the theory of Doppler cooled trapped ions [Ste86, Esc03, Cir92, Mor97], the laser cooling to the trapped motional ground state was demonstrated with primary Doppler cooling to the mK range and finally with sideband cooling to temperatures on the order of several μK [Die89, Win87, Pei99, Mon95a, Roo99]. The ground state cooling technique is extended to linear ion crystals, where several degrees of motion are cooled simultaneously [Kin98, Roh01].

For the discussion of Doppler and sideband cooling techniques the electronic level scheme of a single trapped ion is simplified to an effective two-level quantum system, providing the qubit levels $|0\rangle$ and $|1\rangle$ analoguous to a spin-1/2 particle. In this experimental work the energy splitting $\hbar\omega_{qs}$ of the qubit levels covers the optical range for the quadrupole qubit system with $|0\rangle = |S_{1/2}, m = 1/2\rangle$ and $|1\rangle = |D_{5/2}, m = 5/2\rangle$. An energy splitting of several kHz is obtained for the Raman qubit $|0\rangle = |S_{1/2}, m = 1/2\rangle$ and $|1\rangle = |S_{1/2}, m = -1/2\rangle$ as Zeeman sublevels of the ground state.

The Hamiltonian H_{st} for the free single ion as a two-level system is completed by the ion interaction with the harmonic pseudopotential of the ion trap, characterized by the secular frequency ω. In a one-dimensional quantum mechanical description with the creation (annihilation) operators $a^\dagger$ (a) [Coh99] for the coordinate $x = \sqrt{\hbar/2m\omega}\,(a^\dagger + a)$ and the momentum $p = i\sqrt{2m\omega/\hbar}\,(a^\dagger - a)$ of the trapped ion, the Hamiltonian H_{st} follows to

$$H_{st} = \frac{1}{2}\,\hbar\omega_{qs}\,\sigma_z + \hbar\omega\left(a^\dagger a + \frac{1}{2}\right).$$

The Hamiltonian H_{in} describing the trapped ion-light interaction is based on a laser at frequency ω_l with $E(t) = E_0 \cdot \cos(kx - \omega_l t + \phi)$ as an one-dimensional electric field. Using the rotating wave approximation, the Hamiltonian H_{in} is expressed in terms of the creation and annihilation operators, the coupling strength Ω and the Lamb-Dicke parameter η [Roo00] using $\sigma^+ = |1\rangle\langle 0|$ and $\sigma^- = |0\rangle\langle 1|$ to

$$H_{in} = \frac{1}{2}\hbar\Omega \left(e^{i\eta(a^\dagger + a)}\sigma^+ e^{-i\omega_l t} + e^{-i\eta(a^\dagger + a)}\sigma^- e^{i\omega_l t} \right) .$$

The Lamb-Dicke parameter $\eta = k\,z_0$ describes the coupling cross section by the effective wavenumber $k = \omega_l/c \cdot \cos(\tilde{k}, \tilde{x})$ multiplied with the lateral spread of the ions zero-point wavefunction $z_0 = \sqrt{\hbar/2m\omega}$ confined in the harmonic potential. The Rabi frequency Ω describes the effective coupling strength for the qubit transition, depending on the polarization and amplitude of the electromagnetic field $\tilde{E}$ projected on the position operator $\tilde{r}$ and the atomic transition properties. The Rabi frequency is determined by $\Omega = \langle 0|\,(e\tilde{r}\cdot\tilde{E})\,(\tilde{r}\cdot\tilde{k})\,|1\rangle\,/4\hbar$ for a quadrupole transition.

Following [Roo00], the Hamiltonian $H = H_{st} + H_{in}$ of the trapped ion interacting with the laser is calculated using $U = e^{iH_{st}t/\hbar}$ in the interaction picture $\tilde{H} = U^\dagger H U$ to

$$\tilde{H} = \frac{1}{2}\hbar\Omega \left(e^{i\eta(\tilde{a}^\dagger + \tilde{a})}\sigma^+ e^{-i(\omega_l - \omega_{qs})t} + e^{-i\eta(\tilde{a}^\dagger + \tilde{a})}\sigma^- e^{i(\omega_l - \omega_{qs})t} \right)$$

with $\tilde{a} = a\,e^{i\omega t}$ respectively. The qubit levels $|0\rangle$ and $|1\rangle$ of the free ion are coupled by the laser to the motion of the trapped ion in the harmonic potential with the vibrational eigenstates $|n\rangle$. The transition $|0, n\rangle \leftrightarrow |1, n+m\rangle$ driven by the laser at a frequency $\omega_l = \omega_{qs} + m\omega$ couples the qubit states with the vibrational eigenstates of the single trapped ion. The laser frequency ω_l is modulated by the axial trap frequency ω from the ions point of view, so the state $|\psi(t)\rangle$ as a general solution of the time-dependent Schrödinger equation $i\hbar\partial_t|\psi(t)\rangle = \tilde{H}|\psi(t)\rangle$ is described as

$$|\psi(t)\rangle = \sum_n c_{0,n}(t)|0, n\rangle + c_{1,n}(t)|1, n\rangle .$$

If the trapped ion, characterized by its spatial extension at the vibrational quantum number n, is confined to a insignificant fraction of the laser wavelength, the qubit transition combined with an additional vibrational excitation is limited in the Lamb-Dicke regime to three distinct transitions based on $m = 0, \pm 1$. The Lamb-Dicke regime is characterized by the Lamb-Dicke limit $\eta^2(2n + 1) \ll 1$, where the recoil frequency $\omega_{re} = \hbar k^2/2m$ is significantly smaller $\omega_{re} \ll \omega$ than the axial frequency ω [Esc03].

In the Lamb-Dicke regime three transitions are fundamental, the optical carrier transition $|0, n\rangle \leftrightarrow |1, n\rangle$ without motional coupling and two sideband transitions with the creation $|0, n\rangle \leftrightarrow |1, n+1\rangle$ and annihilation $|0, n\rangle \leftrightarrow |1, n-1\rangle$ of a single phonon (Fig. 4.1). The carrier transition is characterized by $\omega_l = \omega_{qs}$ with a coupling strength of $\Omega = \Omega_{n,n}(1 - \eta^2 n)$. The interaction Hamiltonian $\tilde{H}$ is reduced to $\tilde{H} = \hbar\Omega_{n,n}(\sigma^+ + \sigma^-)$. Opposite to the unchanged motional state of the trapped ion at the carrier transition, the photon absorption resonantly on the red sideband at $\omega_l = \omega_{qs} - \omega$ reduces the phonon population by a single quanta with a reduced coupling strength of $\Omega_{n,n-1} = \Omega\eta\sqrt{n}$. The interaction Hamiltonian is determined to the Jaynes-Cummings Hamiltonian $\tilde{H} = i\hbar\Omega_{n,n-1}(\tilde{a}\sigma^+ - \tilde{a}^\dagger\sigma^-)$, which describes the interaction of a two-level atom with a single electromagnetic field mode. The motional state of the single trapped ion is changed by an additional phonon if the laser is tuned to the blue sideband at $\omega_l = \omega_{qs} + \omega$. The coupling strength $\Omega_{n,n+1} = \Omega\eta\sqrt{n+1}$ is on the same order compared to the red sideband, but is clearly distinguishable for phonon populations of a few quanta. The resulting anti-Jaynes Cummings Hamiltonian is denoted as $\tilde{H} = i\hbar\Omega_{n,n+1}(\tilde{a}^\dagger\sigma^+ - \tilde{a}\sigma^-)$. Higher order red and blue sidebands refer to higher order processes and are still existing, but suppressed significantly in the deep Lamb-Dicke regime.

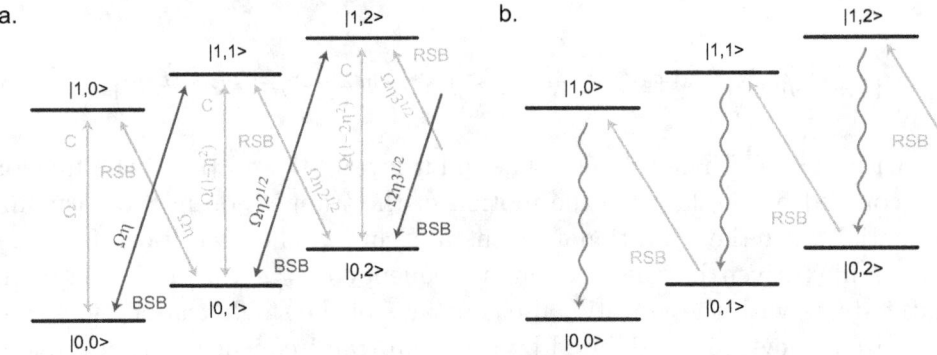

Figure 4.1: Level scheme for the qubit states $|0\rangle$ and $|1\rangle$ coupled to discrete motional states $|n\rangle$ of a single trapped ion: (a) In the Lamb-Dicke regime only the carrier (C) and the first lower (RSB) and upper sideband (BSB) transitions are significant. The coupling strength is shown for the three lowest motional states $|n = 0\rangle$ to $|n = 2\rangle$. The carrier transition is decreasing from 1 at $|n = 0\rangle$, the coupling of the sideband increases strongly for $|n > 0\rangle$ (cp. Fig. 7.7a). (b) A simple technical scheme of the sideband cooling technique illustrates the excitation on the first red motional sideband (RSB) followed by an fluorescence on the carrier transition leading to a phonon annihilation of a single quanta.

The coupling strength of the carrier and sideband transitions is different inside the Lamb-Dicke regime compared to the outer regime. Measuring the coherent coupling of both qubit levels $|0\rangle$ and $|1\rangle$ for the fundamental carrier and first order red and blue sideband transitions the coherent population transfer depends on the pulse duration Δt to

$$P(|0,n\rangle \leftrightarrow |1,m\rangle) = \frac{1}{2}\left(1 - e^{-\gamma\Delta t}\sum_n P_n\cos(2\Omega_{n,m}\Delta t)\right),$$

using a thermal state distribution $P_n = \bar{n}^n/(1+\bar{n})^{n+1}$ with an estimated mean phonon number $\bar{n}$ (Fig. 4.2). The decoherence of the coherent manipulation can be considered by the additional empirical exponential decay $e^{-\gamma \Delta t}$. Typically the coherence time $1/\gamma$ is on the order of a couple of hundreds of μs. The population transfer in the Lamb-Dicke regime at mean vibrational quantum numbers on the order of $\bar{n} = 10$ shows strong coupling on the carrier transition and a weaker coupling of the red and blue sideband (Fig. 7.6). Outside the Lamb-Dicke regime, the coupling strength Ω of the carrier is decreased, while the coupling on both sidebands is increased strongly (Fig. 7.7) - Rabi oscillations on the sidebands at mean phonon numbers $\bar{n} \approx 100$ occur.

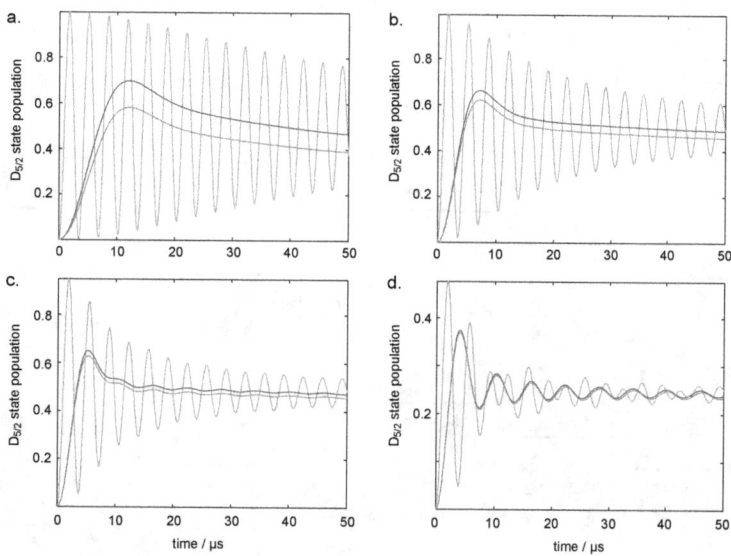

Figure 4.2: Theoretical Rabi oscillations of a thermal state for different mean phonon numbers $\bar{n}$: (a) $\bar{n} = 5$, (b) $\bar{n} = 15$, (c) $\bar{n} = 30$, (d) $\bar{n} = 150$. The carrier (gray) and the upper (blue) and lower (red) motional sideband are calculated based on realistic parameters used in the experiment: For a single ^{40}Ca$^+$ ion confined with a axial frequency of $\omega_{ax} = (2\pi) \, 1.5$MHz and a 729nm laser tilted 45° to the linear trap axis, a coupling strength of $\Omega_{0,0} = (2\pi) \, 150$kHz represented by the power of the 729nm laser is estimated. The axial Lamb-Dicke parameter follows to $\eta = 0.056$. For increasing mean phonon numbers $\bar{n}$ Rabi oscillations on the sidebands appear.

The sideband resolved detection in the Lamb-Dicke regime is crucial for the quantum state engineering of the effective two-level atomic system. The coherent quantum state manipulation on the carrier transition is implemented for single qubit rotations, the controlled excitation of the first motional sidebands allows to establish two qubit quantum gates. Even the selective addressing of the red sideband is fundamental for cooling to the motional ground state.

4.2   Laser cooling techniques

Several cooling techniques for trapped ions in Paul and Penning traps are demonstrated experimentally so far [Ita95]. Particularly with regard to Paul traps, besides Doppler laser cooling by inelastic photon scattering [Die89] and sympathetic cooling by Coulomb interaction [Roh01, Bar03], the damping of the ion motion can be realized even by electromechanical cooling with feedback control using trap electrode voltages [Bus06]. Based on this initial cooling techniques a mean phonon population of several phonons is obtained, that allows further cooling below the Lamb-Dicke limit to the motional ground state using sideband cooling. In this experimental setup the trapped single ion is Doppler precooled on a closed transition to a mean phonon number of approximately $\bar{n} \approx 10$ according to a temperature of several mK, and then sideband cooled using the quadrupole transition with a remaining temperature of some μK.

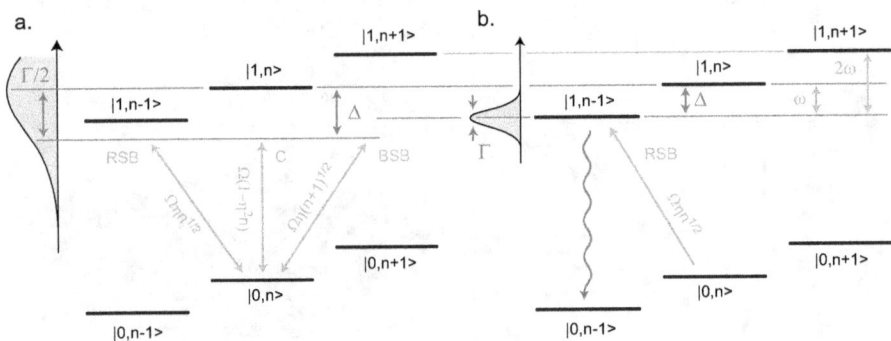

Figure 4.3: Laser cooling in the Lamb-Dicke regime $\eta\sqrt{n} \ll 1$, the spatial extension of the ions motional wave packet is much smaller than the laser wavelength. Three transitions with m = 0, ±1 are provided, the others are suppressed. (a) Doppler cooling is realized in the regime of weak confinement $\Gamma \gg \omega$. The sidebands are not resolved in this regime. The cooling laser is detuned to $\Delta = -\Gamma/2$. (b) Sideband cooling is achieved on a narrow transition in the strong binding regime $\Gamma \ll \omega$. The motional sidebands are resolved. The laser is detuned by $\Delta = -\omega$ to the resonance of the red sideband transition.

Two different cooling regimes [Ste86] are covered by Doppler precooling and sideband cooling of a trapped single ion, corresponding to a weak ($\Gamma \gg \omega$) and strong confinement ($\Gamma \ll \omega$) in the Lamb-Dicke regime (Fig. 4.3). Starting with Doppler cooling, the decay rate Γ of the dipole cooling transition, corresponding to the natural linewidth by $\tau = 1/\Gamma$, is much larger than the axial motional frequency $\omega \ll \Gamma$ of the trapped ion. This allows that the photon absorption and fluorescence is considered as an instantaneous process, so the momentum change is similar to a free atom cooled with a frequency-shifted laser. In the following sideband cooling regime below the Lamb-Dicke limit, the decay rate Γ of the quadrupole cooling transition is much smaller than the axial motional frequency $\omega \gg \Gamma$. The ion is confined spatially to a fraction of the cooling laser frequency. The motional sidebands caused by oscillation of the ion are resolved by the cooling laser. Particular sidebands are adressible without any influence of the adjacent carrier or higher order sideband transitions.

4.2.1 Doppler cooling

The Doppler cooling for a $^{40}\text{Ca}^+$ ion is realized using the dipole transition $S_{1/2} \leftrightarrow P_{1/2}$. The final temperature for Doppler cooling $T = \hbar\Gamma/2k_B$ is determined by the decay rate of the cooling transition and is limited for $\Gamma = 1/7.7\text{ns}$ to $T = 0.5\text{mK}$. The mean energy of the ion $E = k_B T$ is quantized using a mean phonon number $\bar{n}$ to $E = \hbar\omega(\bar{n} + 1/2)$, so the minimal mean phonon number for a two-level system in axial direction

$$\bar{n} = \frac{1}{2}\left(\frac{\Gamma}{\omega} - 1\right)$$

depends on the transition rate Γ and the strength ω of the axial confinement for the trapped ion. For an axial motional frequency of $\omega = (2\pi)\,1\text{MHz}$ and a transition rate near 130MHz, which are used in the experimental setup of the microchip trap, the Doppler cooling leads to a mean phonon number of $\bar{n} = 9.8$. A strong axial confinement produced by miniaturizing the trap electrode geometry is needed as a promising starting point for the sideband cooling on the $S_{1/2} \leftrightarrow D_{5/2}$ transition: The oscillation of the ion at axial frequencies of 500kHz to 200kHz leads to mean phonon numbers of approximately 20 to 52 quanta, located near the Lamb-Dicke limit and beyond for sideband cooling at the quadrupole transition.

4.2.2 Sideband cooling

The subsequent sideband cooling on the atomic transition $S_{1/2} \leftrightarrow D_{5/2}$ requires a well-resolved sideband spectrum by the cooling laser. A strong axial confinement, combined with a decay rate enhanced significantly compared to the axial motional frequency, is crucial for operating in the Lamb-Dicke regime. A 729nm laser linewidth of a fraction of the axial motional frequency addresses the motional sidebands and avoids a perturbing interaction with adjacent sidebands or the carrier. The spectral intensity of the 729nm laser determines the cooling rate directly, so a frequency stabilization to a small linewidth benefits the cooling. For the sequence of sideband cooling to reduce the motional quanta step-by-step the 729nm cooling laser is tuned to the first vibrational red sideband. Each excitation on the red sideband at

$\omega_{qs} - \omega$, followed by a spontaneous emission on the carrier transition, reduces the phonon population by a single vibrational quanta. The ion is cooled successively to the motional ground state, which is decoupled from further excitation by the laser tuned to the red sideband.

Considering the Lamb-Dicke parameter η, the spatial spread $\sqrt{\hbar/2m\omega}$ of the zero order vibrational wavefunction $|n = 0\rangle$ for a single $^{40}Ca^+$ ion is enlarged to 11nm, 16nm and 25nm for motional frequencies of 1MHz, 500KHz and 200KHz respectively. The Lamb-Dicke parameter η is calculated to 0.068, 0.097 and 0.153 using a 729nm laser, tilted by 45° to the axis of vibrational motion. The constraint for operating below the Lamb-Dicke limit is fulfilled strongly for an axial motional frequency of 1MHz only, leading to $\eta^2(2\bar{n} + 1) = 0.096 \ll 1$. At an axial confinement of 500kHz the ion is located in the intermediate region $\eta^2(2\bar{n} + 1) = 0.39$, the efficiency of resolved sideband cooling is decreased (Fig. 4.4). The condition collapses at 200KHz with $\eta^2(2\bar{n}+1) = 2.42$, illustrating the difference between "trapped" ions with $\eta \ll 1$ and "quasi-free" ions at $\eta > 1$. The significance of an axial well-confined ion is shown clearly by these parameters, to allow cooling on the red sideband preparing the ion in the axial motional ground state for further operations.

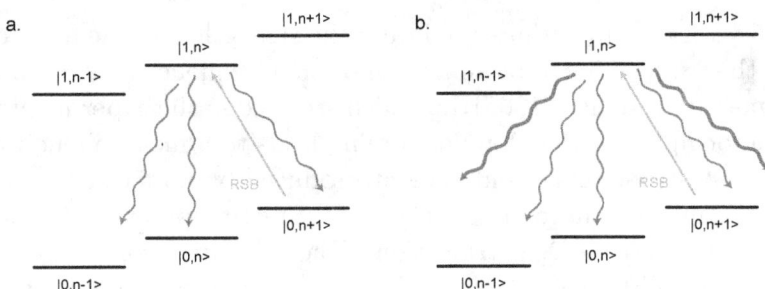

Figure 4.4: Sideband cooling inside (a) and outside (b) of the Lamb-Dicke regime: The strong diffusion of the population prevents efficient cooling beyond the Lamb-Dicke regime. More sophisticated cooling schemes in the intermediate range of $\eta = 1\ldots3$ are discussed by [Mor97], where a first confinement pulse with a large red detuning and subsequent ground state pumping on the blue sideband are proposed theoretically to cool a single ion to its motional ground state.

The Doppler and sideband cooling in the linear ion trap (Fig. 4.5) is discussed for one-dimensional cooling of the motion parallel to the linear trap axis. The cooling techniques can be extended without limitations to the three-dimensional case, depending on the axial and secular motional frequencies. Then three Lamb-Dicke factors η are existent, and the efficiency of Doppler cooling is divided into the components of the cooling laser direction relative to the principal axes of motion. The sideband cooling can be realized by alternating cooling on the first lower motional sidebands for the axial mode and both of the radial modes [Pei99]. Successively the ion is cooled in all three directions of motion. In idealized linear traps the radial motion is decoupled from the axial vibrational motion used as a quantum bus. In special cases, anharmonicities or an unfortunate choice of parameters may couple radial motional quanta to residual axial motion [Alh96, Alh95], producing decoherence effects on the coherent manipulation of the ions quantum state. Especially for microscaled trap designs with electrode shapes predetermined by the microfabrication technique various anharmonicities of the electric trapping potential can occur, but can be reduced efficiently by numerical optimization of the trap geometry.

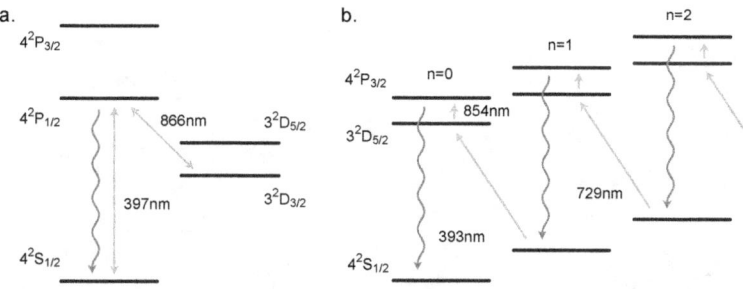

Figure 4.5: Transitions for Doppler and sideband cooling of $^{40}Ca^+$ [Mar94]: (a) The Doppler cooling is realized using a fast decaying dipole transition $S_{1/2} \leftrightarrow P_{1/2}$, even used for fluorescence detection at 397nm of the trapped ions. A repumper laser at 866nm for the transition $P_{1/2} \leftrightarrow D_{3/2}$ is required. (b) The quadrupole transition $S_{1/2} \leftrightarrow D_{5/2}$ is used for sideband cooling. A depletion laser at 854nm for the transition $D_{5/2} \leftrightarrow P_{3/2}$ excites the ion to the fast decaying $P_{3/2}$ state, which is connected to the ground state $S_{1/2}$ with a dipole transition to assure an adequate cooling rate.

Scalable ion quantum computing is based on the interacting of several ions trapped in the same axial potential. The quantum state manipulation is performed addressing vibrational modes of the linear crystal as a quantum bus. A general approach for the application of scalable quantum algorithms using multi-segmented microtraps is based on the manipulation of a two-ion linear crystal in a separated axial potential. A linear crystal out of two ions shows two different axial eigenmodes [Jam98], the center-of-mass (COM) and the stretch mode at the doubled resonance frequency compared to the COM mode. Beyond the Lamb-Dicke limit a separation of the different modes is not possible, but a Doppler precooling to the Lamb-Dicke regime allows a resolved sideband cooling of each individual eigenmode [Mor01]. In the Lamb-Dicke regime the first lower and upper sideband of each eigenfrequency is existing beside the carrier transition, so the ion crystal can be cooled individually to the Doppler limit or the motional ground state.

Chapter 5

Microtrap fabrication

The microtraps used in the experiments with ions or microparticles are linear Paul traps [Pau90] with segmented control electrodes on a microscopic scale. Multiple control electrodes provide an efficient shuttling of ions between independent trap regions. The combination of shuttling operations and the partitioning of the trap in adjacent independent controllable trap regions is a significant improvement for scalable quantum information processing. The scalability of the microtraps is fundamental for the fabrication methods of large-scaled quantum devices [Ste97, Ste06a].

The multi-layer microtrap (5.1), the microparticle trap (5.2) and the planar microtraps (5.3) are manufactured using different fabrication techniques. The capability of each production method covers the UHV compatibility, the dimensionality of the electrode geometry, the surface quality of the trap electrodes, the geometrical accuracy and the option of a monolithic fabrication. In addition the electrical characteristics like breakdown voltage and electrical conductivity of the electrodes, just as permittivity and loss tangent of the dielectric substrates for the radiofrequency are significant trap parameters. Furthermore the ion-insulator spacing is a crucial design parameter for stable trap operation. The applicability of the microtrap for operation in a cryogenic environment is a substantive feature for future experiments [Lab08a, Ant08].

The multi-layer microtrap is based on a three-dimensional electrode geometry fabricated with laser micromachining using femtosecond laser pulses. The several layers are assembled manually. The trapped ions are confined with electrodes directed to the ion from all three degrees of freedom. The microparticle trap is a five-layer planar trap fabricated with standard printed circuit board (PCB) technology. Similar in the trap design the fabrication of the planar microtraps is lithography based with single-layer microscopic trap electrodes. The ion confinement is achieved with an open-electrode geometry providing an open half space.

5.1 Multi-layer microchip trap

The fundamental design of the scalable multi-layer microtrap is borrowed from the electrode geometry of a conventional linear Paul mass filter [Pau53, Pau55]. The microfabricated ion trap (Fig. 5.1) is based on a two-layer electrode design modeled as a logical continuation of scaled down Paul mass filters [Sym98, Fre99] with additional multi-segmented dc electrodes.

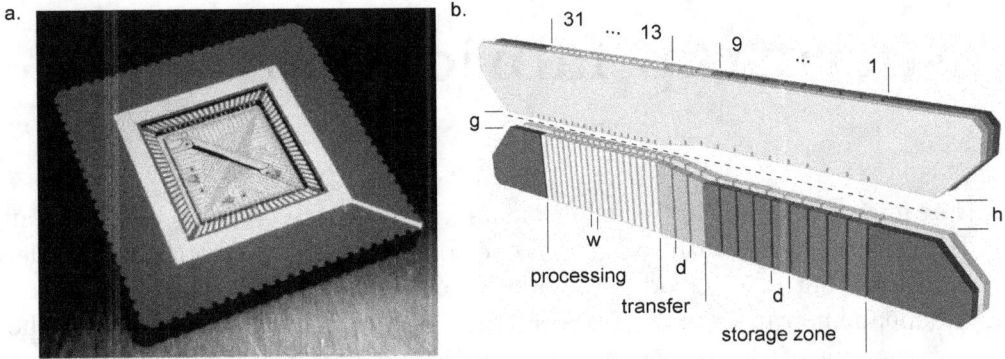

Figure 5.1: Linear multi-segmented microtrap: (a) All rf and dc electrodes are controlled independently. (b) Scheme of the trap: 31 electrode pairs of three different regions for ion storage, transfer and quantum information processing are characterized by the slit widths g and h and the electrode dimensions w and d. The overall linear trap length amounts to 5.8mm.

Each layer of the trap contains an unsegmented rf bar and a dc electrode divided into 31 independent controllable adjacent segments. The rf and dc electrodes of each layer are separated by an elongated slit parallel to these electrodes for optical fluorescence detection and laser access to the trapped ions. Both electrode layers are as symmetric as possible for well-balanced electric fields on the trap axis (Fig. 5.1a). The geometrical symmetry axis is named as 'linear trap axis'. Two independent electrode layers are separated by an insulation layer (Fig. 5.1b), so the cross section of the linear trap shows a three-layer stack design. The trapped ions are confined between both electrode layers. The three-dimensional electrode design ensures an electromagnetic shielding for the ions concerning external electric fields. The insulation layer acts as a spacer and, in future, allows the integration of additional optical elements, such as optical fibers (Fig. 10.1), that are well-aligned to the trapped ions right away.

The microtrap is assembled in a UHV-compatible ceramic chip carrier with the advantage of simple electrical connectivity and fast exchange. The ceramic chip carrier allows the integration in UHV-compatible electronic boards with low-pass filters and, in future, with digital electronic components close to the trap like fast digital-to-analog converters and radiofrequency amplifiers inside the vacuum.

Each layer of the three-layer stack is manufactured separately out of polished Al_2O_3 wafers[1] using laser micromachining and electron beam coating. The fabrication process (Fig. 5.2) starts with laser cutting[2] of the $125\mu m$ thick blank wafers using a femtosecond pulsed laser source. The central slit and the dc electrode fingers will be machined (Fig. 5.2a), the width of the slit taper from h = $500\mu m$ to g = $250\mu m$. The width of the dc electrode fingers changes from d = $250\mu m$ to w = $100\mu m$ in the narrower region. The adjacent fingers have a constant length of $200\mu m$ and are separated by a spacing of $30\mu m$. The trenches in the rf bar oppositely to the dc electrode gaps provide the axial field symmetry, the length of these rf notches is $60\mu m$ with a spacing of $30\mu m$. Additional holes with a diameter of $240\mu m$ allow the mutual alignment of the three layers and provide the mounting of the trap. The feed holes for positioning have their counterpart on all layers at the same place, the mounting holes are one-sided only.

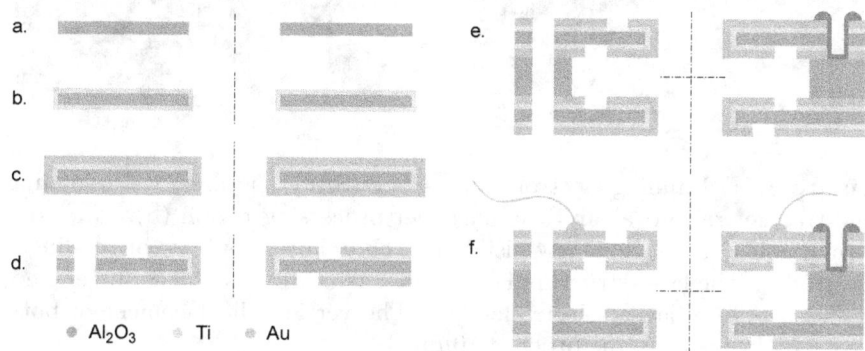

Figure 5.2: Fabrication and assembly of the microtrap: the cross section shows the processing of a single Al_2O_3 wafer. It includes laser cutting (a), cleaning and the metal deposition of the adhesive layer (b) and the gold layer (c). Then the electrodes are created using laser ablation (d). The assembly starts with the mounting of the wafer stack with adhesive (e) and is finalized by bonding the trap inside the chip carrier (f). The dashed cross indicates the rf node and therefore the ions position.

The electron beam coating of the laser-cutted Al_2O_3 wafers is done after an ultrasonic cleaning procedure using acetone (C_3H_6O), isopropyl (C_2H_6O) and Caro's acid (H_2SO_5). The preparation for the coating is finalized by an oxygen plasma cleaning. The blank wafer is coated under continuous rotation with a declination angle of $45°$ for an overall closed evaporation layer. The thickness of the titanium adhesive layer is 50nm (Fig. 5.2b), then a 500nm thick gold layer (Fig. 5.2c) follows the first metallization. The metal deposition is done for each side of the wafer separately. The surface roughness is better than 10nm, which was verified by atomic force microscopy.

---

[1]Reinhardt Microtech AG, Wangs, Switzerland
[2]Micreon GmbH, Hannover, Germany

The second laser machining step structures the wafer metallization to generate electrically independent electrodes, the bond pads and the alignment holes (Fig. 5.2d). Finally, the wafer is laser diced to its top or bottom layer dimensions (Fig. 11.1). The microtrap consists out of two coated electrode layers and a blank spacer. Successively the top and the bottom layer are aligned with spicules manually and mounted on the intermediate spacer using UHV compatible epoxy adhesive (Fig. 5.2e). The epoxy Epo-Tek OG142-13[3] is cured with the uv lamp BlueWave 50[4] within several seconds.

a.
b.

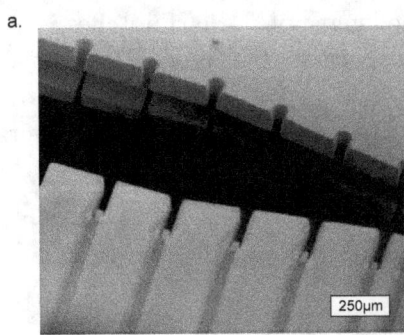
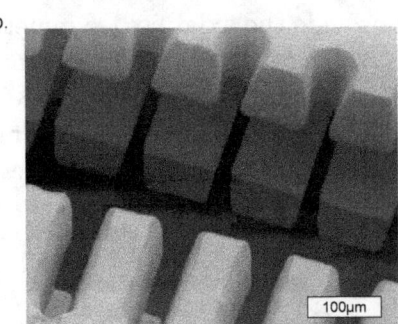

Figure 5.3: Scanning electron microscope (SEM) pictures of the loading and transfer region (a) and the narrower processing region (b). The conformal gold coating is clearly visible by the uniform contrast of the finger electrodes. The laser-structured electrode gap started at the dc finger electrodes defines adjacent electrodes (a). The vertical displacement of both electrode layers is on the order of 10μm (b).

The stack of the bottom electrode layer, the spacer and the top electrode layer is mounted in a center holed UHV compatible leadless ceramic chip carrier from Kyocera[5]. The chip carrier LCC8447001 has 84 contact pins with an outer dimension of 30mm squared and an inner cavity size of 12mm. The microtrap is connected electrically with 15μm diameter gold wire using the ball bonder HB06[6] (Fig. 5.2f). Each dc segment is bonded twice to a contact pad of the chip carrier, the rf electrodes are connected via 12 wires to the chip carrier. All 62 independent dc segments and the rf electrodes are accessible through the chip carrier.

The multiple segment design of the microtrap requires a scalable contact scheme to the chip carrier that is different for each electrode of one pair (Fig. 11.2): The bottom electrode layer has a squared shape. The bond wire and the electrode area visible for the trapped ions are located at the same side. The top layer is more sophisticated technically, because the ball bon-

---

[3]Polytec PT GmbH, Waldbronn, Germany
[4]Dymax Europe GmbH, Frankfurt, Germany
[5]Minitron Elektronik GmbH, Ingolstadt, Germany
[6]TPT GbR, Karlsfeld, Germany

ding and the ions are located on opposite sides. Here, a conformal coating is absolutely essential for the operation of the microtrap (Fig. 5.3a). The trap design avoids blank Al_2O_3 areas because of electric potential disturbances induced by patch charges on dielectric areas. The unavoidable isolation between the electrode segments is shifted away more than $200\mu m$ from the ions position by the extended electrode finger design. Electron microscopy imaging shows an all-sided gold plating of the finger electrodes (Fig. 5.3b).

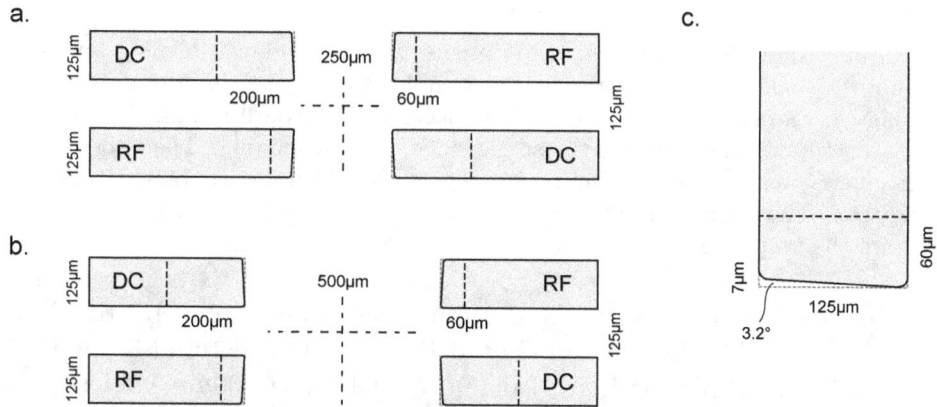

Figure 5.4: Measured radial trap dimensions: cross sections of the processing region (a), loading region (b) and the electrode finger geometry (c) are shown. The cross section dimensions (a) and (b) are reproduced with very high accuracy, the rounded shape of the finger electrodes with a flattening of 3.2° is due to the spatial laser focus during the laser machining (c).

Because of the production process and the manual alignment of the electrode layers some deviations from the ideal geometry occur (Fig. 5.4): The spatial focus of the ablation laser is reflected in the slightly conical shape of the finger electrodes. The laser machining is one-sided only, so the electrode shape from the ions point of view is always the same. The electrode layers of the microtrap used in the experiments so far are aligned manually, the most significant variance is a displacement parallel to the rf bars. A tilting and shift rectangular to the elongated trap axis is negligible.

Deviations from the ideal electrode geometry induce slightly variations of the trapping potentials. Any tilting error of the electrode layers is most critial, but easiest to avoid. The axial displacement error is the most common deviation (Fig. 5.5), the three-dimensional alignment is difficult because of the optical parallax. The axial displacement of the dc segments of an electrode pair broadens the axial potential shape and decreases the axial motional frequency. Second order effects induce marginal rf heating parallel to the axial trap direction close to tapered regions.

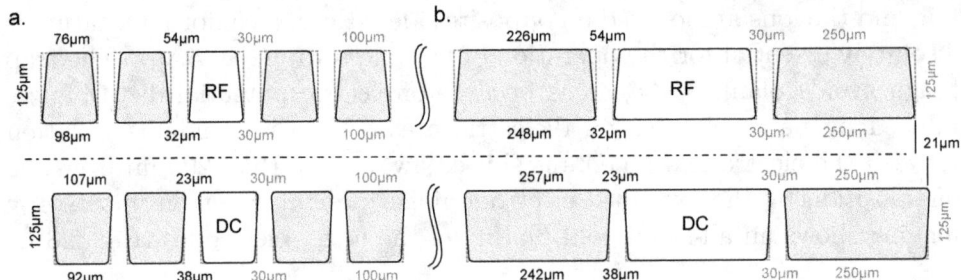

Figure 5.5: Measured axial trap dimensions: minor deviations are apparent from the ideal geometry of the assembled two-layer trap in the processing (a) and loading region (b). The electrode edges are rounded slightly at the outer side because of the single-sided laser machining. The electrode geometry near the linear trap axis is reproduced with an accuracy of a few micron. The main deviation is the lateral displacement error of 21μm issued from the trap assembly.

The microtrap chip carrier is soldered on a printed circuit board (PCB) and installed inside the vacuum (Fig. 6.4). The PCB material is the UHV compatible polyimide laminate Isola P97[7], which is 200μm gold electroplated. Close to the chip carrier each dc electrode is rc low pass filtered based on ceramic SMD 0402 parts with a cutoff frequency of 10MHz. Resistors of 15Ω and capacitors of 1nF were used. The electrical connection to the vacuum feedthroughs is supplied by Kapton ribbon cables with ceramic D-type connectors[8], which are mounted on the PCB with UHV compatible solder.

[7]Isola GmbH, Düren, Germany

[8]allectra GmbH, Berlin, Germany

5.2 Planar multi-layer microparticle trap

The microparticle trap is a proof-of-concept experiment and designed as a surface electrode ion trap (Fig. 5.6) developed for the demonstration of ion shuttling operations. The splitting of ion chains and the deterministic transport of single microparticles above bifurcations of the trap geometry is proven experimentally.

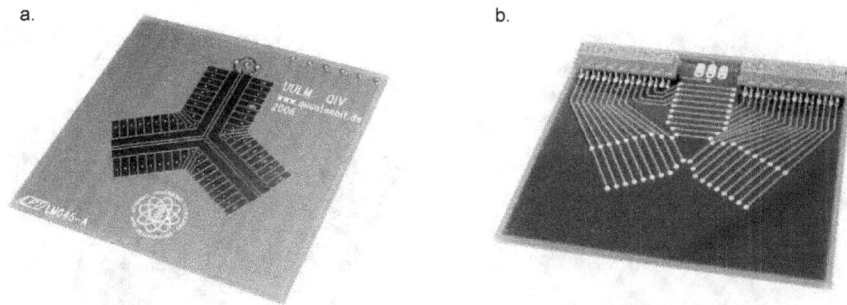

Figure 5.6: Planar microparticle trap: (a) The trap layout is based on a planar linear trap using a Y-shaped three-folded symmetry. The top layer shows the copper trap electrodes with a symmetrical cross section. The wood-like color aside from the trap electrodes is caused by the intermediate ground copper layer. (b) The electrical connectors are surface mounted on the bottom layer, the multiple control electrodes are connected pairwise.

The trap design is based on a planar electrode geometry fabricated with standard PCB technology. The original concept of planar traps using microparticles [Pea06] is extended for scalability to a multi-layer design. The trap geometry consists out of three linear trap regions that are connected at a single bifurcation point. The top electrode layer, a middle continuous ground layer and a bottom layer for electrical connections are separated by intermediate insulation layers of standard FR4 material. All conductor paths are made out of copper. The ground layer allows a complete electrical shielding from the top to the bottom layer, both layers are connected with vias through the ground layer. It is the first planar trap concept which shows a realistic scalable multi-layer surface trap design.

Both ac bars are crossconnected, in each bifurcation the dc electrodes are segmented in 10 electrode pairs. Together with the unsegmented middle electrode the unit cell of the planar trap is controlled by 31 dc electrodes in total. The cross section of the trap design is symmetrically and bounded by 5mm broad longitudinal segmented dc electrodes. The width of the solid ac electrodes is 2mm, the middle dc electrode is unsegmented with a width of 0.4mm. All electrodes are spaced by a constant gap of 0.25mm. The length of the segmented dc electrodes in axial direction amounts to 1.5mm.

The extended hexagon version of the trap (Fig. 5.7) shows clearly the scalability of this concept - the elementary cell of the Y-shaped trap is arranged to a hexagon. Each of the 180 dc electrode pairs of the hexagon trap are multiplexed with their equivalent electrodes at each of the elementary cells, so 31 control voltages are adequate to supply all dc electrodes. Round trips of the trapped ions are possible for the first time - with a scalable design based on the hexagonal crystal structure of graphite.

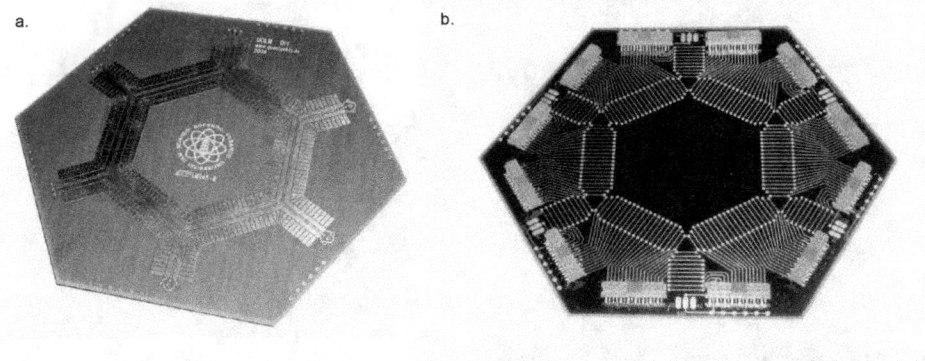

Figure 5.7: Extended hexagon version of the microparticle trap: (a) The fabrication is identical to the scheme used for the elementary cell trap design. The round trip shuttling of ion chains is feasible with this trap design using 31 computer controlled electrode pairs. (b) Every second electrical connector is multiplexed to supply all 180 dc electrode segment pairs with 30 independent voltages only.

Because of the smaller specific mass of the ionized microparticles compared to single ionized atoms, the trap drive frequency is several orders of magnitude lower. The drive frequency is shifted from the radiofrequency range of several 10MHz for ionized atoms to an ac range of several 100Hz for ionized microparticles. The trap geometry determines the required ac and dc peak voltages directly. For the mm-scaled electrode dimensions of this planar trap the ac peak voltage varies from 100V to 850V, the dc electrodes are supplied with a maximum positive voltage of 150V. The smaller the trap dimension, the maximum voltages will decrease further.

A self-developed control unit generate the ac trap drive voltage and 31 dc voltages for ion shuttling (Fig. 5.8): The electronics allows the variation of the trap drive frequency up to 1000Hz continuously. The time-dependent trap voltage is generated using a simplified electronic circuit and is of rectangular instead of sinusoidal shape [Ric73]. Because of the rectangular waveform the peak voltage is increased by a factor of 1.3 compared to the common sinus voltage, because of the first coefficient of the Fourier series of a time-dependent rectangular voltage [Ric75]. A disadvantage is the excitation of higher harmonics in the motion of the trapped microparticle, that cause instabilities for specific trap parameters. The rectangular shaped trap drive voltage is generated by a power transistor BUY71, that switches with a constant frequency between the positive and negative output of two high voltage print modules[9] with opposite polarity. The peak voltage of the high voltage print modules is programmed by a 16-bit digital-to-analog converter MAX541[10] with SPI-compatible interface, the frequency is set by a voltage-to-frequency converter TC9401[11] combined with a MAX514 chip to trigger the switching time of the power transistor. Using the control unit the ac trap drive voltage is programmable with 16-bit resolution in peak voltage up to 1000V and frequencies up to 1kHz.

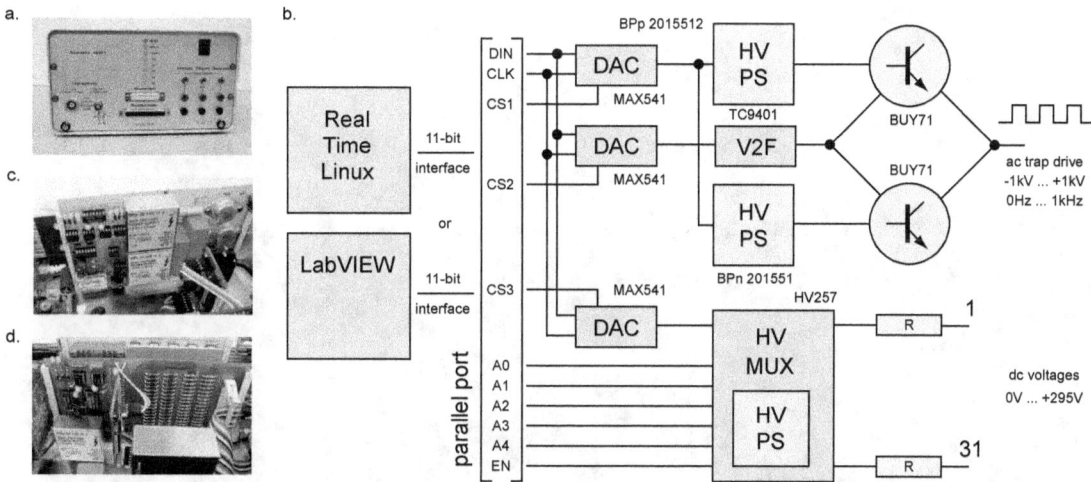

Figure 5.8: (a) Control electronics for the planar microparticle trap. (b) The circuitry illustrates the control of the ac trap frequency and voltage by two 16-bit digital-to-analog converters. The 31 dc voltages are set by just another 16-bit digital-to-analog converter combined with a multichannel high-voltage sample and hold array. The printed circuit boards for the ac trap drive (c) and the dc voltage control (d) show the high level of enhanced integration of the high voltage electronics.

The 32 dc voltages for ion shuttling are generated in a range of 0V to 295V by a third high voltage print module and controlled by the high voltage sample and hold array HV257[12]. The channels are programmable using a 5-bit digital parallel interface and the voltage is set by a 16-bit MAX541 digital-to-analog converter. For fast shuttling operations the slew rate of the dc voltage channels is about 2.5V/μs per channel.

[9]iseg Spezialelektronik GmbH, Radeberg, Germany
[10]Maxim Integrated Products Inc., Sunnyvale, USA
[11]Microchip Technology Inc., Chandler, USA
[12]Supertex Inc., Sunnyvale, USA

All trap parameters are computer controlled with a 11-bit digital interface. A bus of 6-bit is required for the selection of the dc channels, an additional 5-bit wide interface is used for the serial programming of the three digital-to-analog-converters. The digital lines are interfaced to a computer using the parallel port or a high speed digital output card. At the parallel port an update rate per channel of 1kSPS was realized with a LabVIEW control software, limited by the normal process scheduler of the operating system. A real time Linux increases the update rate to 10kSPS per channel, however the update rate is limited by the control system. The theoretical update rate limited by the electronic hardware is in the range of several 100kSPS.

5.3 Planar single-layer microchip trap

The planar microtraps used in the experiment are microfabricated surface traps with a single electrode layer (Fig. 5.9). The concept of a Paul mass filter [Pau53, Pau55] with a two-dimensional cross section [Gee05] is projected on an one-dimensional single electrode layer [Chi05]. The cross section of the linear trap shows two rf electrodes enclosed by segmented control electrodes with a solid dc electrode at the center (Fig. 5.9a) - a segmented middle electrode would require a multi-layer fabrication process. An equal width of the rf bars defines a symmetric planar trap with the principal axes of radial motion parallel and perpendicular to the electrode layer [Bro07]. Tilting of the principal axes is achieved by asymmetric rf electrodes at the cross section [Sei06]. This effects the Doppler cooling of the trapped ion in particular.

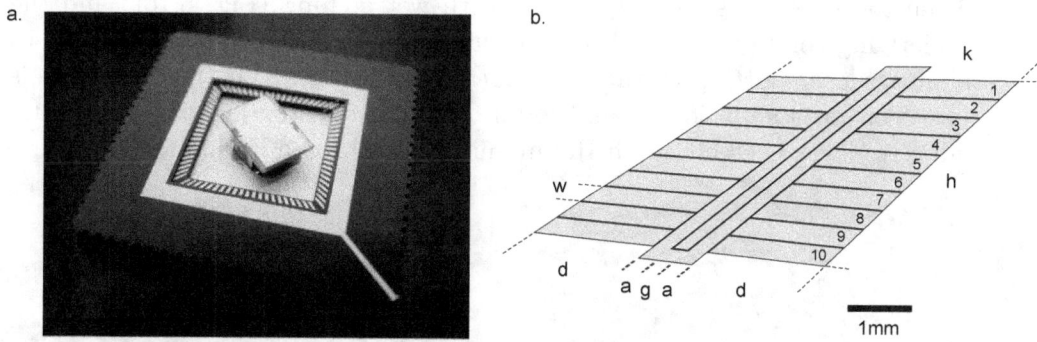

Figure 5.9: Linear planar microtrap: (a) The dc control electrodes and the rf trap electrodes are supplied by the chip carrier contacts. The microtrap is mounted on a metallic nut 3mm above the chip carrier for laser access parallel to the electrodes close to the surface. (b) Scheme of the trap: 10 control electrode pairs with an axial length w and width d enclose the rf bars and the middle dc electrode with the cross section dimensions a and g. The overall horizontal length k is about 4mm, the overall axial trap size amounts to 5mm (microstructured trap supplied by H. Häffner).

A linear symmetric design of a planar microtrap with multiple control segments is most suitable for the demonstration of ion shuttling operations. The study of some characteristic effects that appear using planar cross trap designs like slightly induced rf micromotion along the longitudinal trap axis is enabled by the special electrode geometry used here (Fig. 5.9b). The middle dc electrode is enclosed by the rf bars, the longitudinal axis is bounded by the rf connections that cause a non-vanishing fast decaying ponderomotive potential at both ends of the linear microtrap. The width a of the symmetric shaped rf electrodes is 300μm, the width g of the middle dc electrode is 250μm. Pairs of dc control electrodes with a width d of 1980μm and a lateral dimension w of 400μm are connected separately. The spatial separation of the electrodes is 20μm.

The fabrication of the single-layer planar traps is a monolithic process based on metal deposition and photolithography (Fig. 5.10). The linear planar trap[13] with 10 outer dc electrode pairs is fabricated on polished BK7 glass. After ultrasonic cleaning with acetone (C_3H_6O), isopropyl (C_2H_6O), the preparation is finalized by oxygen plasma cleaning (Fig. 5.10a). An 5nm thick adhesive layer of chromium is deposited (Fig. 5.10b) for the evaporation of 150nm gold as a start layer for galvanoplating (Fig. 5.10c). A thick film photoresist with a thickness of 2μm to 20μm is spin coated and exposed with UV light. After wet etching the remaining resist pattern represent the geometry of the electrode gaps (Fig. 5.10d). The shape of the resist walls has a high aspect ratio, different fabriation processes lead to aspect ratios of 2 to 3. The width of the electrode gaps vary between 1μm and 10μm. The bottom gold coating serves as a start layer for the electrodeposition of gold. The galvanoplating is stopped close to the top of the resist walls (Fig. 5.10e). Then the photoresist is stripped off with wet etching (Fig. 5.10f) and the underlying solid gold and chromium layer are removed out of the gaps to elminate the shorting of the electrodes (Fig. 5.10g). The gold etchant is based on potassium iodide and iodine KI/I_2 chemistry and the chromium adhesive layer is etched with Rohm and Haas Chrome Etchant 18[14].

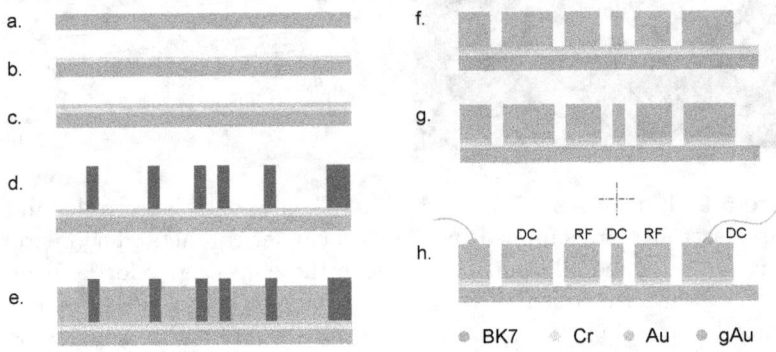

Figure 5.10: Fabrication of the single-layer planar trap: The cross section shows the monolithic processing of the glass wafer. After cleaning (a) a chromium adhesive layer (b) and a thin gold layer (c) are evaporated. The photoresist (d) is patterned and filled up with electrodeposited gold (e). The photoresist (f) together with the conductive start layers (g) are removed. The trap is electrically connected via ball bonding (h). The dashed cross indicates the rf node respectively the ions position above the rf and dc electrodes.

The thickness of the gold galvanoplating determines the aspect ratio of trap electrodes directly. The higher the coating of the gold electrodes, the larger is the distance of the trapped ions to the dielectric areas of the wafer located at the bottom of the electrode gaps. The influence of uneven

[13]Trap chip: H. Häffner, lithography by N. Daniidilis, IQOQI, Innsbruck, Austria
[14]micro resist Technology GmbH, Berlin, Germany

distributed patch potentials to the trapping potentials is reduced with high aspect ratio galvanoplating. The prototype trap used in the experiment has an aspect ratio of about 1 : 4 because of the large electrode gaps, further improvement of the processing should lead to 2 : 1. The gold thickness of the electrode structure is measured using a profilometer with submicron height resolution (Fig. 5.11). The partial cross section of a single electrode indicates a gold layer thickness of 4.69μm. The surface quality is detected by the analysis of the top of the electrode. The height distribution allows the separation of localized defects from the statistical surface variation. The height distribution results in a surface roughness of 40nm.

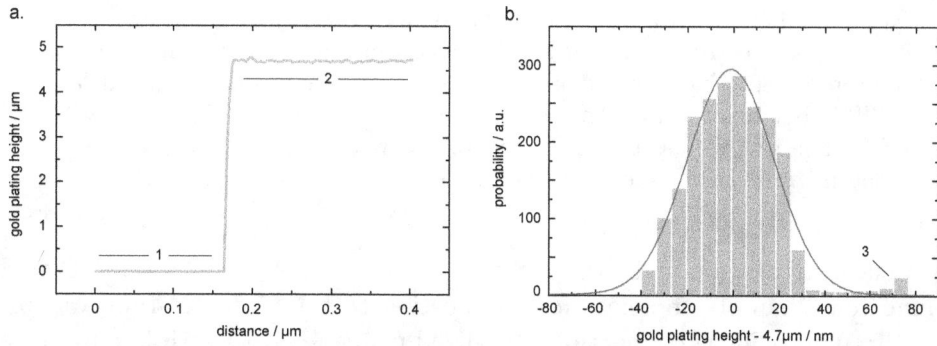

Figure 5.11: Characterization of the gold galvanoplating and the surface roughness: (a) The profile (raw data by N. Daniilidis, IQOQI, Innsbruck, Austria) shows the surface of the glass wafer (1) as a smooth baseline and the surface of an electrode (2). (b) The surface roughness is evaluated using a height distribution on the top of the electrode, excluding localized defects (3) from the statistical surface variation.

The planar trap is assembled in a UHV compatible ceramic chip carrier. The leadless ceramic chip carrier Kyocera LCC8447001[15] has 84 contact pins with a cavity size of 12mm and an outer dimension of 30mm squared. The rf and dc electrodes of the trap are connected with 15μm diameter gold wire using ball bonding[16] to the chip carrier. Each dc electrode is connected via a single bond wire to the carrier contact pads, the rf electrode is contacted using 6 wires in total.

The chip carrier containing the planar trap is soldered on a printed circuit board (PCB). The UHV compatible PCB is made out of polyimide laminate named Isola P97[17] (Fig. 5.12). The copper surface of the one-sided PCB is 200μm gold electroplated. All dc electrodes are rc low pass filtered using ceramic SMD 0402 parts. Resistors of 150Ω and capacitors of 1nF result

---

[15]Minitron Elektronik GmbH, Ingolstadt, Germany

[16]TPT GbR, Karlsfeld, Germany

[17]Isola GmbH, Düren, Germany

a.

b.

Figure 5.12: Electron microscopy and trap assembly: (a) The separation of the gold electrodes is shown using electron microscopy. The lithography process produces well-defined edges with a smooth electrode surface. (b) The assembled trap in the chip carrier is placed on the UHV compatible PCB. The rc low-pass filters are located between the chip carrier and the Kapton wires. Each dc electrode of a pair is controlled individually.

in a cutoff frequency of 1MHz. Kapton ribbon cables with ceramic D-type connectors[18] supply the electrical connection to the 25-pin D-type vacuum feedthroughs. The rf connection is realized using a copper stripline from the vacuum feedthrough to the PCB and is soldered to the electroplated gold surface. It is placed far away from the dc low pass filters to reduce the rf pickup. The rf bars of the planar trap are contacted through the chip carrier pads to the rf voltage supply.

---

[18]allectra GmbH, Berlin, Germany

Chapter 6

Experimental setup

Both ion trap experiments based on the multi-layer and the planar single-layer trap designs are composed out of the same fundamental building blocks. Beyond the ion microtraps the experimental setup of the planar microparticle trap for the demonstration of the shuttling algorithms is completely different.

The five building blocks of the ion trap experiments are summarized as the vacuum system for microtrap operation, laser optics and electronics, detection optics, trap electronics and experiment control: The first building block covers the trap operation and laser access in a UHV environment (6.1) together with the generation of an atomic beam. The lasers for photoionization of the atomic beam, the laser operation and frequency stabilization of the different optical transitions forms the basis for the quantum state spectroscopy (6.2). An efficient fluorescence detection and imaging of crystallized ion chains allows a fast quantum state readout (6.3). The spectral stability and the noise floor of the radiofrequency trap drive and the multi-channel arbitrary waveform generation for the electrode segments are fundamental for the implementation of ion shuttling in quantum information algorithms (6.4). The experimental control system generates the time table for the spectroscopy using different lasers and digital pulse patterns for the ion shuttling electronics (6.5). Additional tasks are the data acquisition and processing of the ion fluorescence.

The main building blocks are unchanged for the operation of the three-dimensional multi-layer and the two-dimensional planar microtraps, however the different trap dimensionality requires modifications of the trap mounting and the optical detection only.

6.1   Trap apparatus

The microtraps are mounted and operated in a modular designed UHV environment with a high pumping capacity and extended optical access (Fig. 6.1). The vacuum housing serves as a base system with pumps, magnetic field coils and windows for laser access and fluorescence detection. A special flange at the top contains all parts necessary for trap operation, so the ex-situ experiment preparation, i.e. alignment of the ovens close to the microtrap, allows a fast change to a different trap on a short timescale.

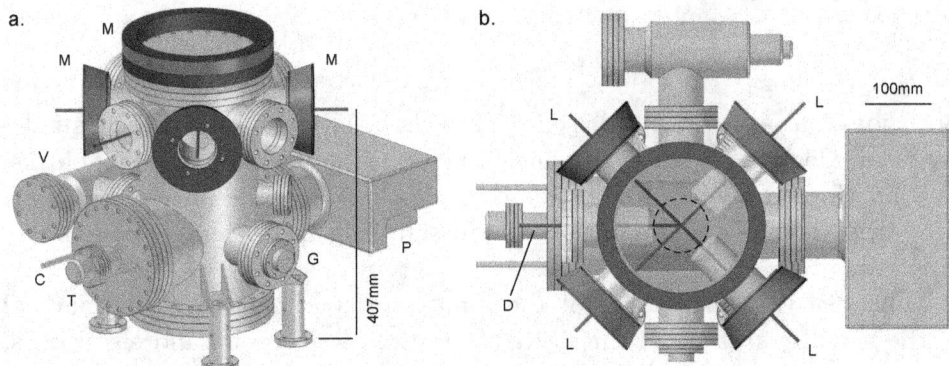

Figure 6.1: Modular trap apparatus: (a) The stainless steel chamber is based on a DN200CF cross section with two different sections: In the lower section the ion pump (P), the titanium pump (T) with cryoshroud (C), the ion gauge (G) and valve (V) are installed, the upper section contains the special trap flange (orange) with the magnetic field coils (red) (M). (b) Optical access for laser (L) and detection (D) is provided by inverted viewports close to the trap center. The exemplary detection geometry corresponds to the multi-layer microtrap.

The trap apparatus is build as a modular system using a special flange integrated all components for trap operation. The mounting of the microtrap, the electrical feedthroughs for the trap drive, the multiple control segments and the current supplies for the atomic beam ovens are installed on the same DN200CF top flange for easy maintenance.

The vacuum base system (Fig. 6.1a) is divided into the upper experimental part with an octagon cross for optical access and the lower technical part with vacuum technique installed. The DN63CF angle valve[1] is used for initial evacuation with a 250l/s turbomolecular pump[2]. A DN63CF 65l/s StarCell ion pump[3] and a 500l/s combination of a titanium subli-

---

[1]VAT Vakuumventile AG, Haag, Swiss
[2]Leybold Vakuum GmbH, Köln, Germany
[3]Varian Inc., Palo Alto, USA

mation pump[4] and a water cooled cryoshroud is operated for closed-cycle pumping. The background pressure is monitored using an extended Bayard-Alpert type UHV gauge[5]. The center of the upper section is located 407mm above the bottom (Fig. 6.1b). Two DN63CF inverted viewports[6] with an inner diameter of 45mm allow perpendicular laser configurations tilted to the microtrap. The optical detection is optimized for a maximal numerical aperture using a third DN63CF inverted viewport close to the trapped ions. The window material of the inverted viewports is fused silica, the DN63CF plane windows[7] installed at the vacuum housing are made out of BK7. The inverted viewports and the plane windows are coated with a custom-made antireflection layer[8] for the optical wavelengths of 397nm and 729nm. The concept of the inverted viewports provides optical access very close to the trap center with the ability of installing large scalable microtraps, illustrated by the dashed circle with a diameter of 85mm only.

For trap operation at UHV conditions a custom-made bakeout tent[9] with a heater fan is used for inital evacuation after trap installation. A maximum temperature of 150°C for a couple of days is required with slow in- and decreasing temperature ramps to achieve an end pressure of 10^{-10}mbar. The end pressure is limited unpredictably because of remaining flux from the soldered SMD parts located at the PCB trap mount. However, a structurally identical trap apparatus for a different experiment is operated at a working pressure of $< 10^{-11}$mbar.

6.1.1 Laser configuration and optical access

The different laser beams for the photoionization of the atoms and the excitation, manipulation and detection of the ions are aligned through the inverted viewports. Most of the laser beams are leaded to the viewports in opposite direction to minimize the stray light by scattering. The two groups of superimposed laser beams are perpendicular to each other and are parallel to the symmetry axis of their pair of magnetic field coils. The laser configuration for both experiments is very similar, while the fluorescence detection at the three-dimensional microtrap experiment is embedded in the horizontal plane (Fig. 6.2), in contrast to the planar trap experiment (Fig. 6.3). There the fluorescence detection is located vertically to the trap surface and is realized with an additional inverted viewport implemented in the trap flange.

---

[4]tectra GmbH, Frankfurt am Main, Germany
[5]Varian Inc., Palo Alto, USA
[6]UKAEA, Special Techniques, Abingdon, United Kingdom
[7]VACOM GmbH, Jena, Germany
[8]Tafelmaier Dünnschicht-Technik GmbH, Rosenheim, Germany
[9]tectra GmbH, Frankfurt am Main, Germany

The three-dimensional microtrap is mounted on the trap flange with the electrode surface parallel to the detection viewport (Fig. 6.2a). The axial trap direction is within the plane defined by the lasers and the optical path for the detection. The fluorescence detection of the trapped ions is perpendicular to the axial trap direction using a separate viewport to suppress stray light detection that worsens the effiency of the quantum state readout. A CCD camera and a photomultiplier are used for photon counting.

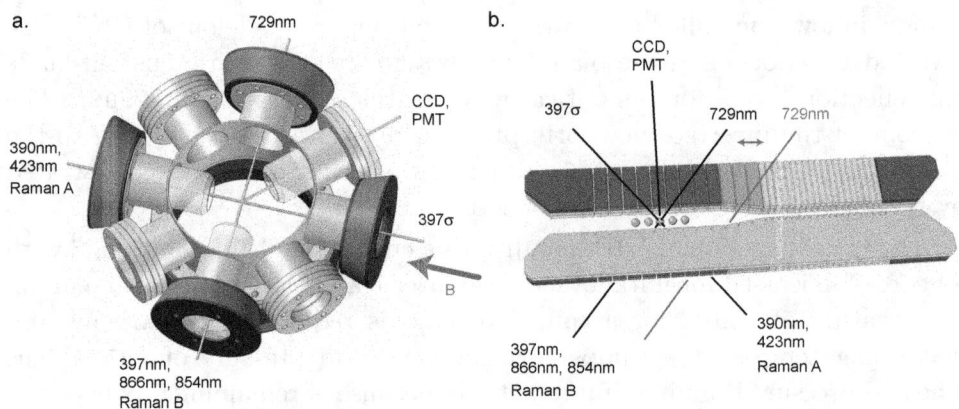

Figure 6.2: Geometry of the laser configuration at the microtrap experiment: (a) The fluorescence detection of the trapped ions is separated by an additional inverted viewport from the laser beams. (b) All laser beams are fixed positioned except the laser at 729nm, that is moved along the axial trap direction for the transport spectroscopy experiments.

The 397nm laser for Doppler cooling, the 866nm laser for repumping and the 854nm laser for level depletion are oriented collinear to the laser at 729nm and perpendicular to the magnetic field B (Fig.6.2b). The 397nm σ-polarized laser for optical pumping is aligned parallel to the magnetic field. The lasers for photoionization at 423nm and 390nm shares the inverted viewport with the Raman laser A, which is perpendicular to the Raman laser B. All lasers are oriented to the axial trap direction with an angle of $\pm45°$ respectively.

This laser configuration shows different orientations for the k-vectors for coherent quantum state manipulation using the laser at 729nm or the Raman laser scheme: While the k-vector of the laser driving the quadrupole transition has non-vanishing components in the axial and radial direction, the effective k-vector for the Raman laser beams is aligned parallel to the axial trap direction. The quantum jump spectroscopy shows the radial and axial sidebands of ion motion using the quadrupole transition with the 729nm laser. The spectroscopy using the Zeeman sublevels of the ground state with the Raman scheme shows a pure axial sideband spectrum with the consequence of non-excitable radial sidebands.

Especially the coherent manipulation of the ions quantum state at the quadrupole transition as well as the interferometrically stable Raman laser beams requires a mechanically stable and well-focused laser setup. The laser optics is installed at an additional threaded alumina breadboard at the same level in front of the viewports. The lasers except the laser at 729nm are focused with a single lens down to a waist size of approximately 30μm at the ion. The laser intensities at the position of the trapped ions are for 397nm about 30μW for Doppler cooling and 0.3mW for detection, for 866nm the intensity is 0.5mW and for 854nm about 0.1mW. The laser beam driving the quadrupole transition with an intensity of 60mW is parallelized with a telescope at first and then focused down to 20μm at the ion. The laser for photoionization are focused down to 150μm at the ion. The laser intensities are 1.5mW at 423nm and 0.8mW at 375nm and lead to a loading rate of a several ions per second depending on the flux of the atomic beam source.

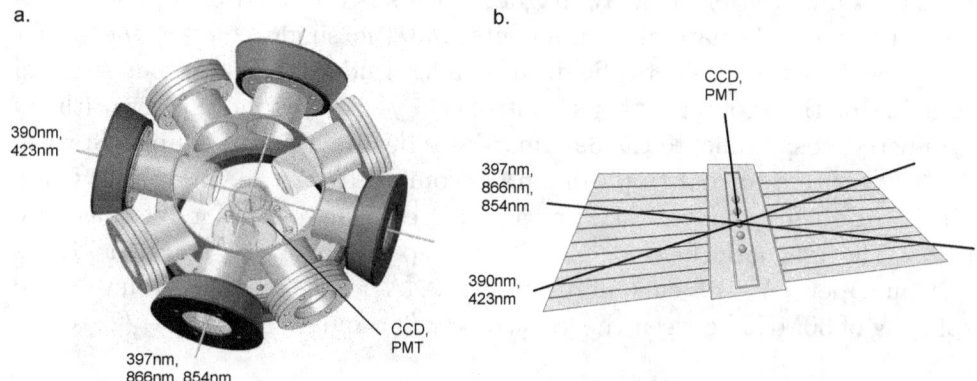

Figure 6.3: Geometry of the laser configuration at the planar trap experiment: (a) The fluorescence detection of the trapped ions is separated from the plane defined by the laser beams to a vertical position above the trap surface. (b) The lasers are aligned to a specific segment pair of the trap. The laser for Doppler cooling is oriented under a non-vanishing angle to both principal trap axes.

The two-dimensional planar trap is mounted on the trap flange with the electrode surface horizontally parallel. The axial trap direction is within the plane defined by the lasers and is tilted to an angle of $\pm45°$ respectively. The fluorescence detection of the trapped ions is perpendicular to the trap surface with a separate inverted viewport implemented in the trap flange (Fig. 6.3a). The detection system and lasers are the same as used in the three-dimensional microtrap setup. The laser systems for photoionization at 375nm and 423nm are aligned collinear, the lasers at 397nm, 866nm and 854nm for Doppler cooling and repumping are aligned perpendicular. The laser beams are aligned approximately 150μm above the surface and coincides with their focus at the trapping electrode segment (Fig. 6.3b).

6.1.2   Magnetic field coils

The magnetic field coils allows the manipulation of the Zeeman splitting of
the ground state sublevels as well as the magnetic field compensation in all
three spatial dimensions. Two pairs of magnetic field coils are operated in a
Helmholtz configuration and orientied perpendicular to each other (Fig.6.1).
The coils are installed at a distance of 23cm at the DN63CF viewports
close to the vacuum vessel. The inner diameter is 10cm, the whorl of the
1.5mm diameter copper wire is agglutinated to the alumina body achieving
a improved thermal conductivity for the exhaust heat. Albeit the Helmholtz
condition - the radius of the coils equals the distance of the coil pair - is not
fulfilled for the Zeeman coils, the magnetic field at the position of the trapped
ions covers a small area of interest of several hundred μm^2 only, where the
field homogenity is preserved adequately. The single upper magnetic field
coil with an inner diameter of 20cm is mounted at the trap flange.

The compensation of stray magnetic fields is obtained using two Helmholtz pairs for the lateral components and the single coil for the upper
component of the magnetic field. The magnitude of the Zeeman sublevel
splitting of the ground state is controlled by the Helmholtz pair with its
symmetry axis parallel to the 397nm σ-laser beam. A power supply (Statron
2225.2[10]) with a current ripple of 2mA is combined with an external low pass
filter (cut-off frequency of 1Hz) as a stable current source for the magnetic
field. The magnetic field strength at the position of the trapped ions is
1.2Gauss per 1A, the current used was 2.4A for a Zeeman ground state
splitting of 30MHz between the $|S_{1/2}, m = 1/2\rangle$ and $|S_{1/2}, m = -1/2\rangle$ state.

6.1.3   Calcium oven

The Calcium ions are produced in a resonant two step photoionization process from an effusive neutral atomic beam, which is directed to the loading
zone of the microtrap (Fig. 6.4a). The oven is placed 10mm apart in front
of the loading zone, so the widening of the beam cross section is minimized
in the free flight distance. This prevents the extensive deposition of neutral Calcium on the trap electrodes, because atomic layers of Calcium oxide
sustain the generation of patch potentials near the trapped ions disturbing
the pseudopotential of the trap.

The oven is made out of a 40mm long stainless steel tube with an inner
diameter of 0.8mm and a wall thickness of 0.2mm. The electric resistively
heated tube is clamped at the posterior end to a 0.7mm diameter stainless
steel rod. At the front side a 10mm wide tantalum stripe[11] with a thickness
of 0.4mm is welded 8mm below the oven orifice and then welded to another

---

[10]Statron Gerätetechnik GmbH, Fürstenwalde, Germany
[11]Goodfellow Cambridge Ltd., Huntingdon, UK

a. 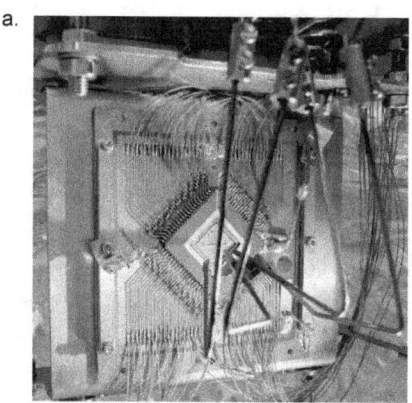 b.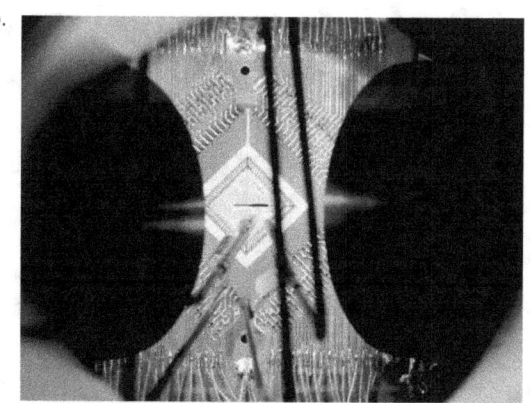

Figure 6.4: Microtrap mounting and oven design: (a) The microtrap is centered vertically and the chip carrier is soldered to a gold-plated printed circuit board with SMD low pass filters close to the trap. Outer contact pads connect wires for the dc electrodes to the microtrap. The solid rf connection is realized with a 1mm copper stripline soldered to the vacuum feedthrough directly. In front of the microtrap two ovens are installed. (b) The ovens for generating the effusive neutral atomic beam are aligned to the loading zone. In the vacuum housing the ovens are delimited by the inverted viewports.

stainless steel rod with the same diameter. The stainless steel rods are bent for alignment to the loading zone and connected to copper electrical feedthroughs for current supply.

The oven is filled up to the end of the tantalum stripe with Calcium granules[12]. The electrical resistance of the cold oven is about 150Ω, during the first operation the Calcium is deoxidized at a current of 4.5A within 6 hours. Because of the limited lifetime of the trapped ions, i.e. due to the ion loss during transport experiments, the continuous operation of the oven is necessary with a permanent current of 3A applied. For single ion loading only a fraction of the mean sublimation temperature of 450°C at a pressure of 10^{-10}mbar is required, the loading rate is varied between one Calcium ion per 10s and 0.5s.

The simple oven design is advantageously regarding to the small effective heat capacity, so the reaction time between a variation of the current to the atomic flux is within seconds. Because of the continuous deposition of neutral Calcium a short-circuit of the outer microtrap electrodes is critical, but the microtrap used in the experiments is up to now in operation for 2.5 years without technical problems. The critical area of about 1mm^2 on the area of the microchip trap is apart from the direct view of the trapped ions, so the influence of deposited neutral Calcium atoms as source of local patch potential is omitted (Fig. 6.4b).

---

[12] Alfa Aesar GmbH, Karlsruhe, Germany

6.2   Diode lasers and laser stabilization

The atomic transitions used for the quantum state manipulation of the ions and even for the photoionization of the neutrals are accessible with diode lasers. All laser systems are commercial devices[13]. The diode laser design is modular and based on a grating stabilized laser diode in a Littrow setup. This master oscillator is combined with a tapered amplifier diode or an external ring resonator for individual purpose at each wavelength.

The master oscillator DL 100 consists out of the laser diode, a grating for Littrow stabilization and a collimator lens in between. The first diffraction order of the blazed grating is focused back to the laser diode, creating a resonator out of the laser diodes rear facet and the external grating. The typical linewidth of the grating stabilized laser is about 1MHz compared to 100MHz of a free-running laser diode. The typical output power is about 5mW. The wavelength is tunable by the injection current of the p-n junction and the ambient temperature. The laser for Doppler cooling and fluorescence detection at 397nm, the repumper at 854nm and the laser for level depletion at 866nm are of this design and will stabilized further by a cavity. Even the free-running laser at 375nm for photoionization is of this type.

For the quadrupole transition at 729nm more laser intensity is needed to drive the dipole forbidden transition efficiently. A tapered amplifier diode is seeded by a master oscillator, which controls the internal modes of the amplifier completely and allows single mode amplification up to 500mW output power. The laser design is called TA-100. Because of the lifetime of the quadrupole transition and the demand of coherent manipulation the laser is stabilized to a narrow linewidth of about 100Hz.

The lasers at 423nm for the first resonant photoionization step and 397nm for driving the Raman transitions are frequency-doubled laser systems. The master oscillator is mode-matched to an external ring resonator, where the fundamental mode is amplified and the frequency-doubled light is generated by second harmonic generation using a crystal inside the cavity. Only at the Raman laser a tapered amplifier diode is placed between the master oscillator and the ring resonator. The output power of the Raman laser TA-SHG 110 is about 150mW, the laser DL-SHG 110 for photoionization emits about 30mW. The frequency of the lasers is stabilized by an internal lock based on the length of the ring resonator.

All laser systems are monitored by a wavelength meter WSU-30[14] simultaneously using a 8-port fiber switch. An absolute accuracy of 30MHz is achieved, the empiric temporal drift is < 3MHz within 2 hours. The principle is based on several Fizeau interferometers with CCD line detection. The calibration is realized with FMTS spectroscopy on Calcium [Ebl07].

---

[13]Toptica AG, Gräfelfing, Germany
[14]High Finesse GmbH, Tübingen, Germany

The laser frequency stabilization in an ion trapping experiment is slightly different from usual locking schemes based on the spectroscopy of neutrals. While the optical transitions for neutrals are accessible in vapor cells, a spectroscopy lock is common. In the case of ionized atoms an analogue locking scheme for the optical transitions is not practicable, so all laser frequencies are locked to cavities or to a doubling ring resonator by a Pound-Drever-Hall scheme. Fundamental requirements to the cavity lock are a broad locking range depending on the resulting spectral bandwith and a low frequency drift which is lower than a fraction of the transition linewidth. The laser bandwith depends mainly on characteristic cavity parameters like the finesse. Frequency drifts by temperature variations or air circulation result in a change of the cavity length, which is minimized by using special material for the mirror spacer with a low temperature expansion coefficient. An active temperature stabilization is not mandatory, but the operation of the cavity in high vacuum is recommended. Then the thermal conduction is reduced efficiently, for the short-term stability temperature changes by heat radiation get more relevant at laser linewidth of < 100Hz.

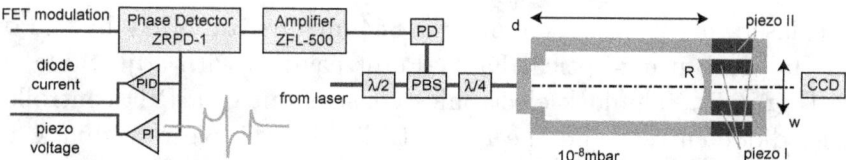

Figure 6.5: Scheme of the low finesse cavity lock: To apply the Pound-Drever-Hall technique the sidebands are modulated directly on the laser using the internal FET. The cavity length d = 100mm with diameter w = 5mm is adjustable by the ring piezos I and II. The cavity mode is monitored by a CCD camera (CCD), the photodiode (PD) in reflection is amplified and mixed for the error signal generation. The lock electronics level the PID-controlled diode current and the PI-controlled piezo voltage.

The Pound-Drever-Hall (PDH) error signal (Fig. 6.5, orange line) is generated by a demodulation of the cavity reflected power from a frequency modulated laser with the modulation frequency itself. The frequency modulation ω_m of several MHz creates sidebands at $\omega \pm \omega_m$ to the carrier frequency ω, which are of opposite phase to each other. The mixing of the photodiode signal demodulates the reflected laser power to dc. The error signal at the carrier frequency ω is nearly linear up to half of the carrier linewidth, which serves as a feedback signal. Slow variations up to 10kHz frequency - limited by its mechanical resonance - are compensated by the piezo of the laser diode grating, fast variations are coupled directly to the laser diode current. The capture range of the PDH lock covers the frequency range of $2\omega_m$ centered at ω because of the sign of the PDH error signal. The PDH locking point of zero is non-sensitive against laser intensity fluctuations, so the laser linewidth is not broadened by this effect.

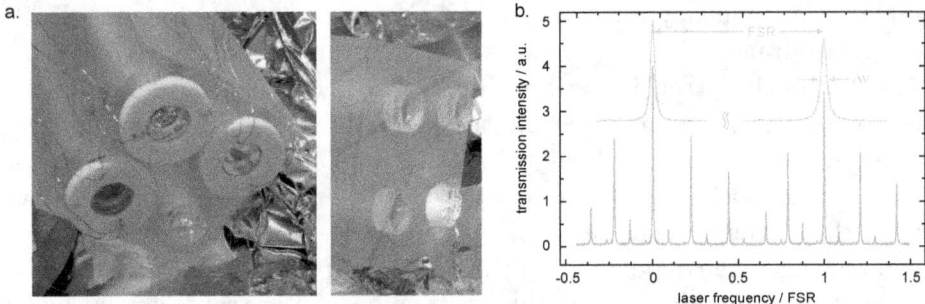

Figure 6.6: Assembly of the low finesse cavity: (a) The cavity base material Zerodur is distinguished by its low temperature expansion coefficient. Three of four cavities are adjustable by length using two independent piezo rings. The glued mirrors are plano-concave on the fixed side (curvature of 250mm) and plano-plano on the moveable side. A reflectivity of 99.1% is achieved by a coating for 397nm and 866nm. (b) Because of the cavity design the transversal modes are separated, the free spectral range (FSR) for the TEM00 mode is 1.5GHz, with a measured finesse of 250.

The lasers at 397nm, 866nm and 854nm are stabilized by a Pound-Drever-Hall lock using optical Fabry-Perot cavities with adjustable cavity length (Fig. 6.5). The lock electronics consists out of a PID controller for the laser diode current and the PI controller for the piezo feedback and is adapted from [Tha99]. The cavities are operated under a vacuum pressure of 10^{-7}mbar to prevent frequency shifts caused by temperature variations. A small 2l/s ion pump[15] is used for continuous pumping (Fig. 6.6). The lasers for repumping at 866nm and level depletion at 854nm are operated at a fixed frequency, while the Doppler cooling laser at 397nm is detuned in a range of several MHz for the optimization of Doppler cooling and micromotion compensation. The detuning of the laser is realized by the variation of the cavity length using the piezo rings. The piezos are made out of PZ27 material[16]. For static adjustment to the atomic resonance a high precision voltage supply EHQ-8010p[17] with 1kV maximum output voltage and a noise level of < 10mV is used. The fine adjustments are implemented with a fast high voltage amplifier miniPiA 103[18] with a maximum voltage of 300V at a ripple of < 50mV. The temporal frequency drift and the resulting laser linewidth depends strongly on the quality of the piezo voltage supplies. The frequency detuning is measured to 10MHz/100V. A laser linewidth of 1kHz is achieved. For the linewidth measurement the laser at 729nm was locked to the low finesse cavity and the linewidth of the carrier transition was detected by quantum jump spectroscopy.

---

[15] Varian Inc., Palo Alto, USA

[16] Ferroperm Piezoceramics, Kvistgaard, Denmark

[17] iseg Spezialelektronik GmbH, Radeberg, Germany

[18] TEM Messtechnik GmbH, Hannover, Germany

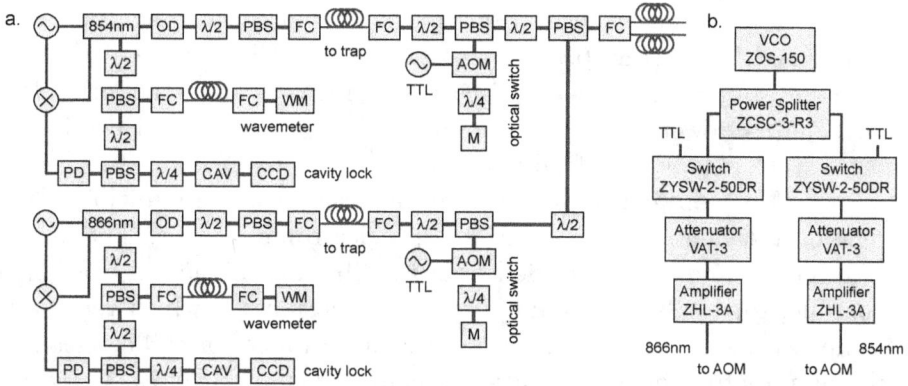

Figure 6.7: Laser optics scheme for the lasers at 866nm and 854nm: (a) The optical path shows the cavity lock and the wavemeter connection. The optical switches are realized in a AOM double pass configuration. The infrared lasers are superimposed for fiber incoupling. (b) The electronics for the optical switches are controlled by digital lines with TTL voltage level.

The infrared lasers at 866nm and 854nm are stabilized to low finesse cavities and connected with a fibre to the wavemeter for coarse frequency measurement (Fig. 6.7a). The optoelectronic shutters allow switching times up to 80ns with an attenuation of 50dB in AOM[19] double path configuration. The radiofrequency supply and digital control for the optical switches are implemented (Fig. 6.7b) using commercial modular parts[20].

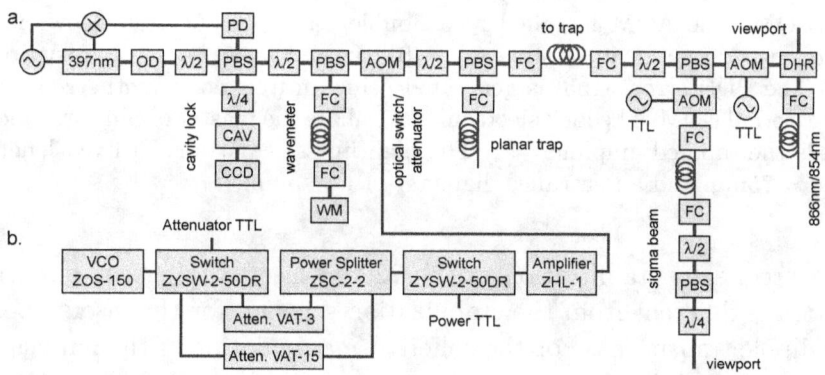

Figure 6.8: Laser optics scheme for the laser at 397nm: (a) The optical path shows the cavity lock, the fibre connection to the wavemeter and the AOM as an optical switch and attenuator (Doppler cooling and detection). The 397nm beam for cooling as well as the σ-beam are digital controllable with digital pulses at TTL voltage levels.

---

[19]Brimrose Corporation, Sparks, USA
[20]Mini-Circuits, Brooklyn, USA

The 397nm laser setup (Fig. 6.8a) is nearly identical to the optical setup of the infrared lasers (Fig. 6.7a), however the AOM is installed in a single pass configuration as an optical switch with an additional attenuation level. The attenuated power is used for the different tasks of Doppler cooling and quantum state readout. A fraction of the 397nm beam is used for optical σ-pumping. Both optical paths are equipped each with an additional AOM in single path configuration used as an optoelectronic shutter. The infrared lasers at 866nm and 854nm guided from the fiber are aligned in front of the trap parallel to the 397nm beam by a dichroitic mirror, which reflects the superimposed beams to a focusing lens on the trapped ions. The electronic circuit for radiofrequency generation, beam attenuation and shuttering is similar to the electronics used for the infrared laser setup (Fig. 6.8b).

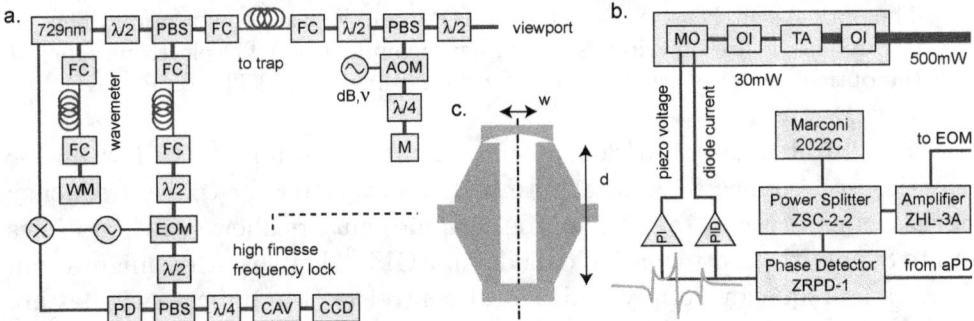

Figure 6.9: Laser optics scheme for the laser at 729nm: (a) The optical path shows the PDH cavity lock with the external laser modulation by an EOM. The AOM installed in a double pass configuration allows fast frequency detuning in a range of ±40MHz and laser intensity variation. (b) The PDH error signal is generated using an frequency synthesizer and an external EOM. The lock electronics stabilizes the master oscillator, which seeds the tapered amplifier. (c) The high finesse cavity has a fixed length of d = 75mm and an entrance diameter of w = 5mm.

The frequency stabilization of the 729nm laser driving the quadrupole transition is different from the stabilization schemes for the lasers operated at the dipole transitions. For the coherent manipulation of the ion using the quadrupole transition an enhanced phase stability is needed, represented by a narrow linewidth of the frequency stabilized laser. The output power of the laser specifies the time period for coherent population transfer directly, combined with the small quadrupole transition matrix element the laser intensity of a tapered amplified system is required.

In the optical path guiding the 729nm laser to the trapped ions an AOM in a double pass configuration is integrated (Fig. 6.9a). A waveform synthesizer VFG-150[21] allows the arbitrary variation of the amplitude, frequency and phase of the radiofrequency signal, which is amplified to drive the AOM with a center frequency of 75MHz. For a coarse wavelength measurement a fraction of the laser intensity is coupled to the wavemeter with a fiber. The frequency stabilization is based on the Pound-Drever-Hall technique with a non-tunable high finesse cavity. For the cavity lock a fraction of the main beam is used with an EOM for beam modulation. The modulation sidebands of the 729nm laser existing solely in the lock branch, therefore the coherent manipulation of the ion is not influenced by the sidebands.

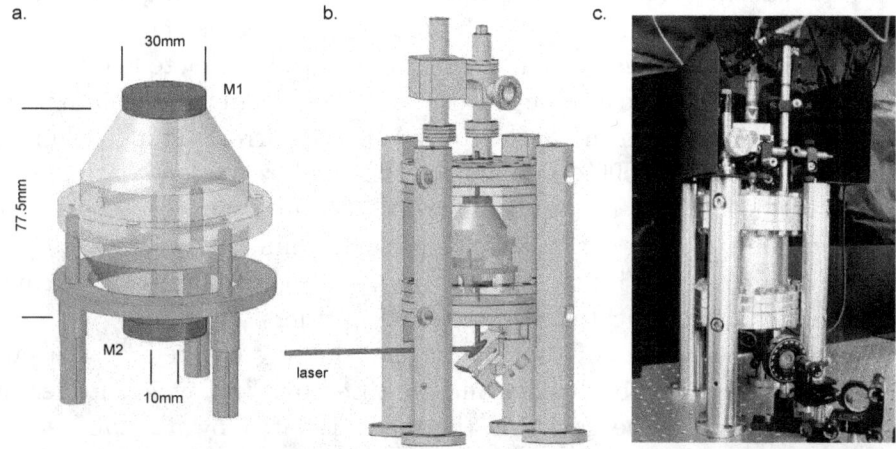

Figure 6.10: Assembly of the high finesse cavity: (a) The Fabry-Perot reference cavity made out of ULE glass is mounted on Teflon rods. (b) The cavity is operated vertically in a special designed vacuum housing at a pressure of 10^{-9}mbar. (c) The cavity is installed on the laser table providing an easy optical access for an upwards directed incoupling of the 729nm laser.

The focus on the frequency stabilization of the 729nm laser is a stable PDH lock circuit providing a spectral linewidth on the order of several 10Hz full width of half maximum. In contrast to the traditional way of a frequency lock with a 1Hz laser linewidth for coherent qubit operations, the 729nm laser is used for qubit manipulation as a starting point for trap characterization only. In the end the qubit will be manipulated by the Raman laser, and the 729nm laser is used for efficient quantum state readout by robust adiabatic passage. The short-term stability is influenced mainly on vibrational effects, the compensation of temperature drifts is essential for long-term frequency stability. Therefore the reference cavity setup is optimized for vibration isolation with a special cavity design and mounting, while an active temperature stabilization is not necessary up to now.

---

[21]Toptica AG, Gräfelfing, Germany

The high finesse cavity is a vertical mid-plane mounted design [Aln08, Lud07, Not05] with a fixed mirror spacing in contrast to the low finesse cavities of tunable length (Fig. 6.10). Because of the transfer of acoustic vibrations by air or mechanical vibrations by contact, the alignment of the cavity mirrors gets worse by vibrational tilting of the mirrors. This results in a reduced finesse and a linewidth broadening. The influence of the horizontal vibrational components are eliminated by a rigid construction, the minimization of the crucial vertical components are considered by a mid-plane vertical mounting (Fig. 6.10a). With a vertical mounting the mirrors are covibrating at a fixed distance. The thermal drifts are compensated roughly by the ULE (ultra low expansion) glass spacer, which has a zero linear thermal expansion coefficient and in addition a vanishing second order part at room temperature. The length of the ULE spacer is 75mm to achieve high mechanical resonance frequencies. The optical contacted mirrors are made of ULE to minimize mechanical stress. The bottom mirror has a plane-plane geometry, the curvature of the concave-plane mirror on top is 50mm. A mirror reflectance of 99.999% is achieved. The high reflection coating exists only at the inner side of the cavity mirrors, the outer surface is layered by an antireflection coating for a wavelength of 729nm. The ULE spacer is mounted with threaded Teflon posts in the vacuum base flange. The cavity was designed[22] and fabricated[23] by external partners.

The optical viewports of the vacuum housing are covered by the same antireflection coating at both sides and are tilted by $2°$ to avoid interference effects (Fig. 6.10b). The cavity is thermally isolated by a small vacuum housing equipped with a Varian $2\,l/s$ ion pump. The 729nm laser is coupled to the cavity from the bottom. The incoupling is optimized for a small optical distance and avoids vibrations using a rigid mounting of the incoupling mirror (Fig. 6.10c). The free spectral range (FSR) of $2\,GHz$ combined with an estimated linewidth of $\sim 10kHz$ results in a finesse of 100000. With this setup the linewidth of the 729nm laser driving the $S_{1/2} \leftrightarrow D_{5/2}$ quadrupole transition is narrowed to $< 100Hz$, which is measured at the carrier transition using quantum jump spectroscopy.

The Raman laser at 397nm can be used as a replacement for the 729nm laser for quantum state manipulation of the Zeeman sublevels $|m = 1/2\rangle$ and $|m = -1/2\rangle$ at the ground state $S_{1/2}$ of $^{40}Ca^+$. The 729nm laser is then used for quantum state readout only. Two phase-coherent Raman beams with a fixed frequency spacing of $3\,GHz$ allow transitions between both Zeeman sublevels. The resulting k-vector of the Raman laser field is oriented parallel to the axial trap direction, because each Raman beam is tilted with an angle of $45°$ to the linear trap axis. The demand of interferometric stability for the Raman beams requires a rigid optomechanical setup.

---

[22] Mark Notcutt, JILA, USA
[23] Advanced Thin Films, Boulder, USA

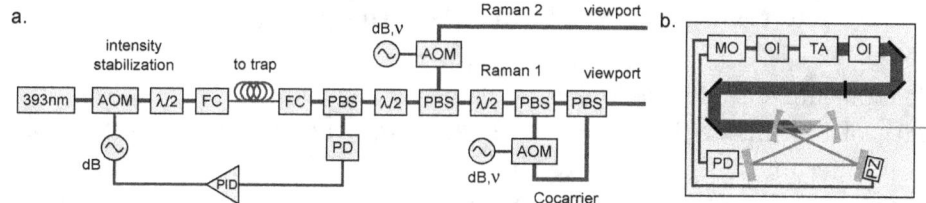

Figure 6.11: Laser optics scheme for the laser at 393nm: (a) The optical path shows the intensity stabilization by an AOM with a photodiode feedback. One Raman beam is tunable in frequency, the other is superimposed with a AOM detuned Raman cocarrier beam. (b) The Raman laser at 393nm consists out of a master oszillator (MO) in Littrow configuration combined with a tapered amplifier diode (TA). Backwards incoupling is prohibited by optical isolators (OI). The light is frequency-doubled by a ring resonator and stabilized using a PDH locking scheme.

The Raman laser intensity is stabilized optionally by a feedback circuit using an AOM and a photodiode (Fig. 6.11a). The Raman beams are created using a polarization beam splitter, an AOM for frequency detuning installed in a single path is used for spectroscopy. The Raman cocarrier is frequency tunable with another AOM and parallel aligned to the remaining beam. Measurements with large Raman detunings on the order of 10 Ghz relative to the $S_{1/2}$ to $P_{3/2}$ transition and Rabi oscillations with several kHz frequency require a high laser output power on the order of 150mW. Therefore a tapered amplifier system with a master oscillator at 794nm is used, which is frequency-doubled using an external ring resonator (Fig. 6.11b).

The bow tie cavity consists out of two confocal and two plain mirrors. The doubling crystal $KNbO_3$ (potassium niobate) is placed in the beam waist between the two confocal mirrors. The laser frequency is stabilized by a variation of the cavity length using a piezo mounted on the plain cavity mirror. The PDH technique is used for generating the error signal for length stabilization, the diode current of the master oscillator is modulated and a photodiode detects the reflected output signal of the ring cavity. The photodiode signal is phase sensitive to the modulated output of the master oscillator and is mixed with the modulation signal. The resulting error signal is the input for the PID controlled lock electronics with feedback to the master oscillator. The mode of the ring cavity output is cleaned by a single-mode fiber placed in between of the intensity stabilization and the optoelectronics for Raman frequency detuning.

6.3   Fluorescence detection and imaging

The quantum state of the ions is detected by their fluorescence at the wavelength of 397nm for $S_{1/2} \leftrightarrow P_{3/2}$. The fluorescence is measured by a photomultiplier and a CCD camera simultaneously. Key parameters of the detection system are the single readout time, the signal-to-noise ratio and the spatial resolution of the imaging. The quantum efficiency of the detection devices combined with the numerical aperture of the lens system results in a fast readout with short exposure times of several ms. The dynamic range of the detector belongs to the signal-to-noise ratio combined with single ion resolution. The technical demand on the lens system together with the CCD camera are the spatial resolution of single ions in an ion chain. The typical distance of adjacent ions is on the order of several μm.

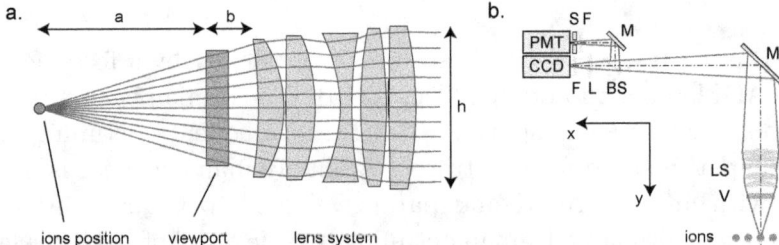

Figure 6.12: Schematics of the optical detection: (a) The custom-made lens is placed b = 17.5mm apart from the viewport, with a remaining distance of a = 42.5mm at the vacuum side. The aperture is h = 38mm. (b) The L-shaped optical path of the detection system consists out of the photomultiplier (PMT), CCD camera (CCD) and the lens (LS) in front of the viewport (V). Some mirrors (M), a beamsplitter (BS) and a lens (L) define the optical path. A band-pass filter (F) for 397nm and a slit (S) are used for optimizing the signal-to-noise ratio of the fluorescence detection.

The trapped ions are centered in the vacuum vessel (Fig. 6.1b). For an optimized collection efficiency of the fluorescence emitted by the ions a lens is placed inside an inverted viewport at the air side (Fig. 6.12a). The custom-made lens system[24] consists out of 5 fused silica lenses, covered with an antireflection coating for light at 397nm and 729nm. The spherical abberation of the 6mm thick fused silica viewport is preconceived in the numerical optimization of the single source imaging at a magnification of 20. The focal length of f = 66.8mm and a source distance of g = 45.1mm result in a image distance of b = 1436.5mm. The numerical aperture of the lens system amounts to 0.27, which limits the optical resolution by Raleigh to 0.85μm. The depth of field is calculated to 18.5μm [Ben08b].

---

[24]Sill Optics GmbH, Wendelstein, Germany

The lens system is the most critical optical component in the detection system, because the readout time is dominated by the collection efficiency of the custom-made lens. The photon collection efficiency is determined by the solid angle $\Delta\Omega$ of the lens . At a working distance of a + b and with the entrance diameter h the solid angle $\Delta\Omega$ follows to

$$\frac{\Delta\Omega}{4\pi} = \frac{1}{2}\left(1 - \sqrt{1 - \frac{1}{1 + (2(a+b)/h)^2}}\right) = 0.0248 \approx \frac{1}{40}$$

The optical detection path (Fig. 6.12b) including the lens system, the photomultiplier and the CCD camera is mounted on a breadboard, which is moveable with a 3-axis translation stage PT3/MBT602[25] (microtrap/planar trap) for focus adjustment. The ion fluorescence is detected simultaneously with the photomultiplier and the CCD camera using a 70:30 beamsplitter. In front of each of the detectors a band-pass filter FF01-377/50-23.7-D[26] transmits the 397nm photons and suppresses infrared light from the 866nm and 854nm lasers. A variable two-dimensional slit[27] minimizes the detection of diffuse reflections from the 397nm laser. The single ion detection is realized easily with a signal-to-noise ratio (SNR) of 4 with a background noise level rate of 4kHz. Further improvement of the slit position allows signal-to-noise ratios of about 10.

The non-cooled photomultiplier P25PC[28] detects 397nm photons with a quantum efficiency of $\eta_{pmt} = 0.27$, the electronics for photon counting is integrated and supports the readout with single digital pulses for single photon counting. The quantum efficiency η_{ccd} of the Peltier-cooled CCD camera iXon 885[29] is increased up to 0.5 using the electron multiplying technique for the CCD chip, while the dark current is decreased efficiently at a chip temperature of $-60°C$. The image area of $256 \cdot 256$ pixels with a pixel size of $8\mu m$ allows the detection of ion chains. The magnification of the optical detection is adapted that at minimum a single CCD pixel remains between adjacent ions of a linear ion string.

The overall detection efficiency η can be calculated using the solid angle $\Delta\Omega/4\pi$ of the lens, the lens transmission loss of 0.96 at 397nm optimized by the antireflection coating and the 397nm filter transmission of 0.88. Including the beamsplitter the total efficiency η can be estimated to 0.004 for the photon counting and 0.003 for the CCD imaging - several hundreds photons are required to trigger a single event, which is sufficient referring to the scattering rate of 10^9 photons per second of the optical dipole transition $S_{1/2} \leftrightarrow P_{3/2}$ for fluorescence detection.

---

[25]Thorlabs, Newton, USA
[26]Semrock, Rochester, USA
[27]Owis GmbH, Staufen, Germany
[28]ET Enterprises, Uxbridge, United Kingdom
[29]Andor, Belfast, Northern Ireland

6.4 Trap voltage supplies

Two types of voltage supplies are required for the operation of the multi-segmented Paul microtraps: A stable radiofrequency voltage supply at a fixed frequency of several tens of MHz with voltages of about $200V_{pp}$ and a programmable multi-channel voltage supply with a range of $-10V$ to $10V$, which is used for the shuttling of ions. The modest requirements on the maximum voltages are predicated upon the microscopic trap dimensions.

6.4.1 Single-channel rf voltage

The radiofrequency voltage in the VHF range for the trap drive is characterized by a high peak voltage compared to a standard signal generator output and a passive impedance matching to the rf trap electrodes. The rf voltage is generated by a Marconi 2019 signal generator[30] followed by a Mini-Circuits LZY-1 amplifier[31] with a fixed gain of 44dB. Optionally the variable attenuator Mini-Circuits ZX73-2500+ is placed in between, providing a voltage controllable attenuation in the range from 5dB to 55dB. The variable attenuator is installed for ion shuttling experiments in the tapered trap region only, where the rf peak voltage is varied during the ion transport. Between the rf amplifier and the rf trap electrodes a helical resonator is installed at the air side on the top of the trap flange, which is used for further rf amplification and impedance matching of the open-ended circuit (Fig. 6.13). A grounded metal shield on the trap flange minimizes the rf pickup for the laboratory electronics.

The helical resonator is a coaxial quarter-wave resonator with an inner conductor wounded to a helix [Mac59]. The length of the copper solenoid is approximately a quarter-wave of the input frequency and the upper end is grounded to the cylindrical copper shield (Fig. 6.13a). The helix is self supported mainly, polystyrene as a low-loss material stabilizes the coil against mechanical vibrations. A probe coupling at the connected end is used for incoupling and tuning to the resonance frequency. At the open end the helix is soldered to the rf electrical feedthrough, which connects the resonator to the trap rf electrodes using a copper stripline on the vacuum side. Between the helical resonator and the vacuum feedthrough a capacitance divider is installed to determine the incoupled rf voltage to the microtrap. The copper cylinder itself is grounded to the vacuum housing.

The resonator is characterized by the unloaded resonance frequency f_0 and the quality factor Q_0 [Zve61]. The electrical energy is stored in the helix, so the quality factor Q_0 is limited by conductor losses of the heli-coil, the outer shielding and dielectric losses. The total inductivity L_{load}

---

[30]Marconi Instruments Ltd., London, United Kingdom.
[31]Mini-Circuits, New York, USA.

and capacitance C_{load} of the capacitance divider, the vacuum feedthrough, the copper wires and the trap electrodes act as a load for the helical resonator and reduce the resonance frequency to $f < f_0$. The final frequency f results in $1/f = 1/f_0 + 1/f_{load}$. The capacitance of the different components influences the load of the resonator in equal parts, mainly the capacitance divider with 5pF and the electrical feedthrough with 6.7pF are slightly more significant than the rf trap electrodes with a capacitance of 2.5pF for the microtrap and 4.6pF for the planar trap (Fig. 6.13b).

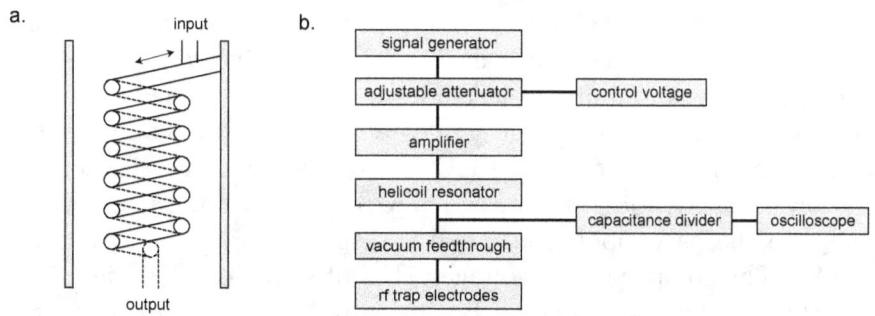

Figure 6.13: Radiofrequency voltage incoupling: (a) Schematics of a helicoil resonator with tunable probe coupling for input and an open-ended output. (b) Functional block diagram of the rf trap voltage supply.

The basic frequency calculation dependent on the resonator dimensions covers the bare frequency f_0 and the unloaded quality factor Q_0 [Mac59]: It is recommended that the diameter D of the shield and the helix diameter d are related to $d/D = 0.5$. The total number of windings N determines the bare frequency $f_0 = 1900/ND$. Additional constraints for a quality factor Q_0 on the order of a couple of hundred are the minimum number of helix turns to $N > 5$ and a helicoil pitch more than twice of the wire radius. The stability of the design against mechanical vibration is crucial to avoid jitter of the resonance frequency and rf amplitude fluctuations respectively.

The helical resonator used for the microtrap is characterized by a bare frequency f_0 of 28MHz, which is reduced to $f = 24.741$MHz with a quality factor of $Q = 80$. The copper helix with a diameter of $d = 4$mm, $N = 4$ turns and a pitch of 8.5mm is integrated in the copper shield with a diameter of $D = 8$cm and a total length of 18cm. The signal generator power of -10dBm results in a trap voltage of $200V_{pp}$ monitored with the capacitance divider. The helical resonator used for the planar trap experiments has a bare resonance frequency of 23MHz and a load resonance frequency of $f = 18.4$MHz with a quality factor of $Q = 120$. The copper shield with its diameter of $D = 8$cm and a length of 25cm contains a 2.5mm diameter helicoil with a pitch of 4mm and 16 turns. An input power of -16dBm at the signal generator is amplified and impedance matched to the rf trap electrodes with a voltage of $180V_{pp}$. The power used is about ~ 400mW.

6.4.2 Multi-channel dc voltages

The operation of multi-segmented Paul traps requires a manifold of programmable stable dc voltages for fast ion transport experiments. Compared to the demonstration of ion shuttling the stationary trapping of an ion chain at a single segment pair is implementable with a single voltage source in principle. This straight concept is used for the demonstration of coherent manipulation of a single ions quantum state including sideband cooling.

The simple concept of a single axial voltage for trapping of a single ion or an ion chain shows the concept for the operation of segmented micro-traps: The axial confinement is realized with a negative voltage $U_n < 0V$ at segment pair n, while all other electrodes $k \neq n$ are grounded with $U_k = 0$. In addition the axial segment electrodes are used for the compensation of radial micromotion with a voltage ΔU. A non-ideal trap fabrication results in a deviation of the radial radiofrequency potential from a perfect symmetric field, so the pseudopotential node is displaced from the point of axial symmetry. The compensation voltage ΔU shifts the ion in radial direction close to the pseudopotential node. Both electrodes of the segment pair n are supplied antisymmetrically with $U_n + \Delta U/2$ and $U_n - \Delta U/2$ to avoid any additional electric potential at the trap center. To preserve the symmetry of the electric potential in axial direction independent of the micromotion compensation, the remaining electrode segments k are supplied with $\Delta U/2$ and $-\Delta U/2$ respectively. The concept is extended to the ion shuttling experiments with a multi-channel voltage source, neglecting position-dependent voltages $\Delta U \approx \Delta U(z)$ for micromotion compensation as lowest-order approximation.

Trapping of ions at a single electrode segment is implemented with a simple analog electronic circuit using operational amplifiers: Two lead acid batteries with a nominal voltage of 12V and 7Ah capacity are connected as a symmetric voltage supply. The negative segment voltage U_n and the micromotion compensation voltage ΔU are adjusted manually by two multi-turn potentiometers working as potential dividers. In an inverting operational amplifier circuit with an OP27[32] the voltage $-\Delta U$ is generated, and with two OP27 summing operational amplifier circuits the voltages $U_n + \Delta U/2$ and $U_n - \Delta U/2$ for each electrode segment of the trapping pair n are obtained. All other electrode pairs k are supplied with ΔU and $-\Delta U$.

The inaccuracies of the output voltages U_n and ΔU are compensated manually with potentiometers, but the gain of the operational amplifier circuits is defined by the ratio of the individual input and output resistors of $1k\Omega$. Therefore the gain error influences the output voltages directly and cause asymmetric errors at the inverting and summing circuitry - these can be minimized by trimmed resistor components. Such asymmetries of the

---

[32]Texas Instruments Inc., Texas, USA.

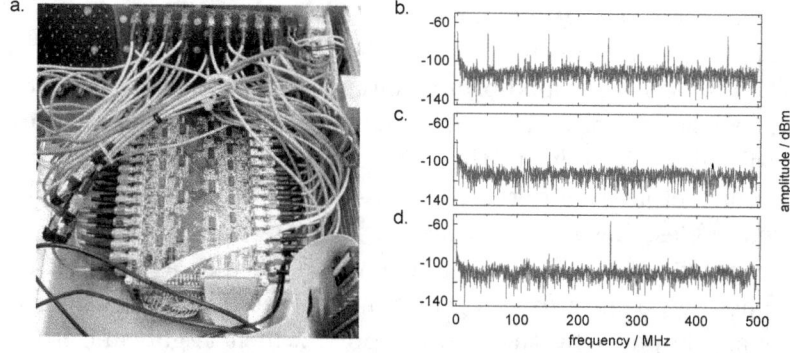

Figure 6.14: Multi-channel voltage supply: (a) Stacked printed circuit boards with 8 digital-to-analog converters TI8814 providing 64 output voltages in total. Noise floor measurement of a NI-6733 analog output (b) at 0V compared to the self-developed TI8814 board at 0V(c). A fourier transformation of a rectangular waveform with an amplitude of 1mV at 256Hz shows the low noise of the TI8814 analog output (d) compared to the noise floor of a NI-6733 card (b).

electrode voltages at a segment pair effect a limited micromotion compensation. In particular arbitrary noise on the voltage supplies induces motional quanta on the ion along the axial direction. For the heating rate measurement as a fundamental attribute of the ion trap it is crucial to avoid any noise-induced motional quanta. The electronic circuitry including the lead acid batteries are installed in metal shielded cases, the output voltages are feeded by shielded cables to the four DB25-type vacuum feedthroughs. The ground of the axial dc voltages is shortened to the radiofrequency ground close to the microtrap chip carrier, which is realized inside the vacuum to prevent any ground loops and is proved empirically.

An extended concept with a manifold of low noise dc control voltages provides transport, splitting and merging operations for ion chains. The Paul microtrap with multiple segments is controlled using multi-channel digital-to-analog converters instead of analog electronic circuits with operational amplifiers. Each electrode of a segment pair is assigned to a specific digital programmable output voltage in a range from −10V to 10V. Because of the sensitive long range Coulomb interaction between the ion and a single electrode potential at least a 16-bit voltage resolution of the full voltage range is required, approximately a voltage step size of 0.3mV. By the way, fast non-adiabatic ion shuttling operations on the order of the trapped ion motional frequencies require a sampling rate of several MSPS per output channel.

The multi-channel voltage source is realized with 16 digital-to-analog converters TI8814[33], each providing 4-channels with a bipolar voltage range

---

[33]Texas Instruments Inc., Texas, USA.

of 10V with 16-bit resolution. The serial SPI-compatible programming interface allows update rates up to 2.5 MSPS per channel at clock frequencies
of 50MHz. Two self-developed printed circuit boards, each carrying 8 digital-
to-analog converters, provide 64 output voltages in total, which are wired
with individual SMA cables to their dedicated pins at 4 DB25-type connectors. The connectors are interfaced with 4 DB25-type vacuum feedthroughs
by shielded cables. Two pairs of lead acid batteries with a voltage of 12V
and 7Ah capacity respectively 6V and 7Ah are used as a symmetric voltage
supply. The digital lines for programming of the TI8814 are interfaced to
the personal computer by the parallel port, which is galvanic isolated via
optocouplers to avoid any pickup of arbitrary noise.

Each digital-to-analog converter TI8814 is interfaced by 4 digital lines -
clock, serial data and two digital inputs for selecting the device and latch.
All 16 devices share the same clock, the remaining 3 digital lines can be
addressed for each device separately. Based on a sophisticated scheme the
control of 16 output voltages with the full sampling rate of 2.5 MSPS in
parallel is achievable: The same single channels of each digital-to-analog
converter are summarized to a subgroup of 16 adjacent electrodes respectively 8 segment pairs. In an ion shuttling experiment the subgroup of
8 segment pairs can be shifted continuously through all electrodes. The
information is processed with the maximal sampling speed, because only a
single channel of each device is addressed at the same time. The demonstration of a non-adiabatic transport operation with a single or a couple of ions
will necessitate this scheme using multi-channel digital-to-analog converters
with a serial programmable interface.

The multi-channel voltage control is interfaced for the adiabatic transport experiments with the parallel port of a personal computer. The parallel
port provides a 8-bit wide digital bus only, so a dedicated line for clock, serial data input and latch is reserved and the 16 remaining digital lines for
selecting of the devices are multiplexed by a 4-bit wide digital bus. The
sampling rate is decreased to 1kSPS, which is adequate for adiabatic shuttling experiments.

In experiments with multi-segmented microfabricated ion traps the programmable voltage source with multiple channels is a crucial device next to
the trap fabrication itself. It is of practical experience to adjust the microtrap dimensions regarding the provided dc voltage range and the expected
motional frequencies of the ions. The allocation of position-dependent voltages for the micromotion compensation is feasible now and the problem of
ion transport is transferred to the development of efficient numerical algorithms.

6.5 Experiment computer control

The requirements for the experiment control are above-average regarding time-critical sequences compared to a standard table-top experiment in quantum optics (Fig. 6.15). All information of the physical system is extracted out of a single ion fluorescence - whether it is or not. The distinction if the metastable state is excited or the ground state is populated is measured statistically with a couple of identical batched experiments for each data point using the electron shelving technique. Each experiment consists out of a timetable, which specifies a digital pattern for switching of the lasers, their frequencies and the time-dependent trap voltages.

All laser beams are switched by the control electronics using digital lines with TTL voltage levels. Two data acquisition devices NI PCI-6733[34] with 16 digital output (DO) channels at a sample rate of 1MSPS are used for measurements with multi-cycle time sequences. The photoionization lasers at 423nm and 375nm are switched by a mechanical shutter for the loading of ions (1DO). The repumper at 866nm, the laser for level depletion at 854nm and the laser at 729nm for coherent quantum state manipulation are controlled independently by optical switches using double pass switchable AOM configurations (3DO). The Doppler cooling laser at 397nm is switched and attenuated by a setup of 2 AOMs (2DO), the optical pumping using the 397σ beam is realized with another AOM switch (1DO). The two Raman laser beams and the Raman co-carrier at 423nm are switched, attenuated and detuned by three independent AOMs in a interferometrically stable optical setup (3DO). The timebase of the optical AOM switches are 10ns approximately depending on the correct laser focus, orders of magnitude faster than the sampling rate of the data acquisition cards and the requirements for the experiment.

Two rf synthesizers RS SML01[35] are used for the variation of the 729nm power, carrier and sideband frequency detuning for spectroscopy. External programming is provided by a GPIB interface. The time duration of the coherent excitation with a rectangular shaped pulse of laser intensity using a timebase of several ns is provided by a electronic delay generator DG535[36] with a GPIB interface. The devices are connected with a GPIB-USB-HS[37] interface to a personal computer. Arbitrary shaped intensity pattern for coherent state manipulation like experiments using robust adiabatic passage are created with a USB interfaced function generator VFG 150[38], which is a more versatile replacement for the delay generator DG535.

---

[34] National Instruments Corporation, Austin, USA.
[35] Rohde & Schwarz GmbH, München, Germany
[36] Stanford Research Systems Inc., Sunnyvale, USA
[37] National Instruments Corporation, Austin, USA.
[38] Toptica Photonics AG, Graefelfing, Germany

The dc voltages for the multiple trap electrodes are supplied by custom-made electronics using TI8814[39] digital-to-analog converters, which are interfaced using a SPI protocol. The programming lines are connected to the parallel port of the personal computer for experiment control. The update rate of the analog channels is limited to 1kHz by the parallel port interface and will be replaced by a FPGA (field programmable gate array) device in the near future to achieve the maximal sampling rate of 2.5MHz for 16 channels simultaneously. The ion transport experiments require a dynamical adjusted rf trap voltage. The variable rf attenuator is connected to a single digital-to-analog channel primary used for the supply of a dc trap voltage. Because of the same timebase the dc trap electrodes and the rf trap supply are triggered for ion shuttling experiments perfectly.

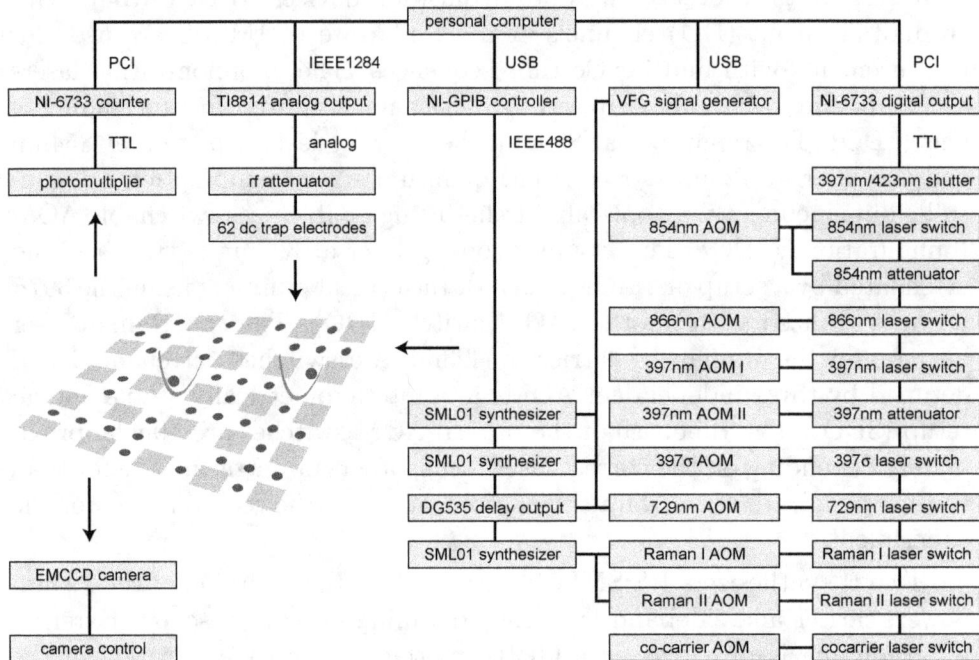

Figure 6.15: Experiment control scheme: The functional block diagram shows the personal computer as the central measurement unit, which controls the experiment and is responsible for the data acquisition by the ions fluorescence. The time-critical task of the digital pattern output of the timetable is assigned to the NI PCI-6733 cards.

The data acquisition of the ions fluorescence is realized with a photomultiplier P25PC[40] connected to the 24-bit counter input channel (1DI) with a timebase of 20MHz of a NI PCI-6733 card.

---

[39]Texas Instruments, Dallas, USA

[40]ET Enterprises Ltd., Uxbridge, United Kingdom

The experiment control is triggered optionally at the edge of the power line using a Schmitt trigger. These minimizes the dephasing of the coherence due to 50Hz magnetic field fluctuations, which is determined directly by the contrast decay of Raman fringes for different delay times (Fig. 7.9). Of course the multi-cycle measurements for each data point are not finished within a fraction of 20ms, but each cycle of different data points is measured at a similar magnetic field, so the averaged coherence is preserved by an order of magnitude.

Almost all measurements are realized by the variation of one parameter, i.e. the 729nm frequency AOM detuning for determing the motional sidebands or the 729nm pulse duration for the observation of Rabi oscillations, and ended with the detection of the binary fluorescence of the single ion and the decision if the ground or metastable quantum state is populated. This statistical measurement is repeated for many times, in general more than 200 times per data point to minimize the statistical error to less than 0.5 percent.

The experiment control software executes timetables with a timebase of $1\mu s$ multiple cycles for each data point. Based on the variation of a single parameter the data acquisition uses the electron shelving technique as the basic detection principle. Each measurement scheme is packetized out of the same modular blocks like i.e. Doppler cooling, sideband cooling, fluorescence detection (Fig. 7.3). It is obvious that a interpreted language with the option of compiled scripts fulfill the requirements at best, so the data analysis framework ROOT developed at CERN is used. The interpreted scripts are programmed in the programming language C/C++ and can be compiled for faster execution. The time critical measurement sequences are generated with ROOT, then downloaded to the NI PCI-6733 card, executed and returned the fluorescence counts from the photomultiplier, which are postprocessed and displayed with GNUPLOT. The timetable is used for the exact laser beam shuttering and the start trigger for the coherent manipulation using the DG535 or the VFG 150. The driver framework Measurement Studio by National Instruments is used for the card programming, with Microsofts Visual C++ the different scripts and the modules used by the driver and the data analysis framework are compiled.

Chapter 7

Microtrap experiments

The experimental results achieved with the multi-layer segmented linear microtrap start with the first trap operation and characterisation of cold linear ion crystals (7.1). The fundamental trap properties are compared with numerical simulations of the electric fields. The coherent quantum state manipulation of a single $^{40}Ca^+$ ion is demonstrated on the quadrupole transition for an optical qubit (7.2).

The microtrap is characterized with quantum jump spectroscopy using the quadrupole transition. The efficient compensation of micromotion is shown (7.2.1), and coherent single ion dynamics (7.3) is initiated by the measurement of Rabi oscillations. The coherence of the quantum state is quantified by Ramsey spectroscopy. A single ion is cooled to the motional ground state using sideband cooling on the quadrupole transition (7.3.1). The applicability of the microtrap for experiments in quantum information science is shown by the measurement of the heating rate in the storage zone (7.3.2). The scalable transport of a single ion allows quantum jump spectroscopy at each trap segment (7.4), demonstrating the combination of spectroscopy and shuttling operations for the first time. The microtrap can be characterized continuously evaluating the motional sidebands at each point of the trap. This method called transport spectroscopy is applied for a complete analysis of the electric field properties in the storage zone and the adjacent tapered trap region.

In contrast to the initialization of the qubit using the quadrupole transition, even the Zeeman sublevels of the ground state are accessible directly via a Raman transition (7.5). Spectroscopy on the Raman transition is realized for the initial tests of a spin qubit.

7.1   Trap operation

All measurements reported here are realized in the storage region of the microtrap. The trap electrode geometry is optimized for the processing of several linear ion crystals each containing a couple of ions. The loading scheme favors the successive trapping of single ions complying to the experimental requirements in quantum information science.

Under typical operating conditions, the trap is supplied with rf voltages of $U = 280V_{pp}$ at a frequency of $\Omega = (2\pi)\,24.841MHz$. The dimensionless stability parameter $q = 2eU/(m\Omega^2)c_2$ results in $q = 0.28$ for the trap electrode geometry of the storage region. The geometric factor $c_2 = 0.52 \cdot 10^7 m^{-2}$ of the quadrupole potential is obtained by numerical simulations of the radial quadrupole field in the storage zone. The frequency of the secular motion $\omega = \Omega \cdot q/2\sqrt{2}$ for the radial confinement is calculated to $\omega = (2\pi)\,1.26MHz$, which is proven experimentally with high accuracy. The trap voltage U can be lowered down to $160V_{pp}$ with remaining stable conditions for single ion trapping, representing a secular frequency of 720kHz.

The successive single ion loading is established using a single electrode pair at the storage region. The adjacent electrode pairs are not involved, both control electrodes are supplied with voltages in the range of $\pm10V$. The ions are trapped at the center of this electrode pair. All adjacent control electrodes are biased with the same voltage for micromotion compensation. The control electrodes at the opposite side are biased with the inverse voltage. The electrode geometry of the storage region allows a tight confinement with an axial motional frequency of 1.2MHz for the control voltage of $-5V$ at the single trapping pair only, which is experimentally verified within 5% accuracy via sideband spectroscopy. The axial voltage can be decreased to $-2V$ without any change of the loading rate, then the axial frequency is shifted down to 480kHz. At weaker axial confinements the successive loading of ion crystals is advantaged, so for single ion experiments discussed here the tight axial confinement is favoured.

The trap is loaded via an effusive atomic calcium beam, which is vaporized using a resistively heated oven. The ofen is operated at a constant current of 2.5A to 3.5A at temperatures in the range of 200°C to 350°C, the beam is directed through the main slit of the microtrap. The effusive beam is ionized partly by isotope selective photoionization using a two step resonant excitation with UV diode lasers [Gul01a]. At first the neutral atoms are excited resonantly from their ground state $4s^1S_0$ to the intermediate state $4p^1P_1$ with a frequency doubled diode laser at 423nm. The second laser at 390nm excites the atom in a Rydberg state with a principal number of $n \sim 30$. Finally the highly excited atoms are ionized by the electric fields of the trap located near the rf node. Both photoionisation lasers are superimposed and focused to the trap center crossing the effusive neutral beam.

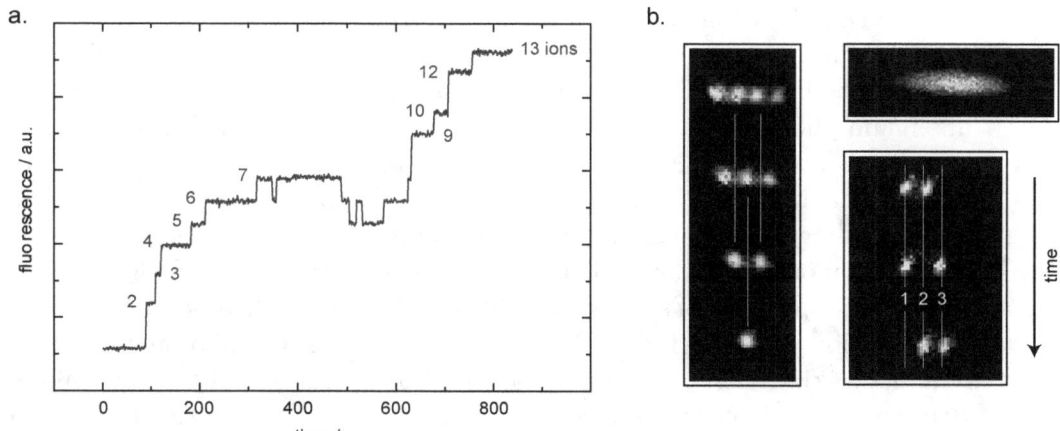

Figure 7.1: Detection and imaging of trapped ^{40}Ca$^+$ ions: (a) The quantized fluorescence signal near 397nm at single ion loading is measured with constant binning of the CCD camera. (b) The fluorescence pictures illustrate the single ion resolution of linear ion crystals (left). A non-crystallized ion cloud is elongated along the axial trap direction (top). The quantum state selective fluorescence is shown at a linear ion crystal out of two ^{40}Ca$^+$ and a dark ion at time-dependent swapping positions (right).

The kinetic energy of a non-cooled ion cloud for trapping is on the order of 20meV to 50meV, combined with heating effects issued from the electric fields of the ion trap an external energy of < 100meV is achieved. Based upon the typical operating conditions of U = 280V$_{pp}$ and Ω = (2π) 24.841MHz, the radial confinement is limited to 0.75eV calculated with the ponderomotive potential approximation $\phi_{ps} = e^2/4m\Omega^2|\nabla\phi|^2$. The axial trap depth results in a couple of eV of the static potential. The mean kinetic energy of the ions shows that the trapped ions has to be generated near the rf node at the trap center. With an approbiate scattering rate of the dipole transition for Doppler cooling the ions are cooled down to the Lamb-Dicke limit successively.

For the detection of the trapped ions the lasers at 397nm and 866nm are essential. Both lasers are crossed with the photoionization lasers and the effusive atomic beam at the rf node. The laser for Doppler cooling and fluorescence detection at 397nm driving the dipole transition S$_{1/2}$ $\leftrightarrow$ P$_{1/2}$ is detuned half of the linewidth $\Gamma/2 \approx$ 11.5MHz to the red side of the resonance for efficient cooling. The initial temperature of the ions leads to a Doppler-broadened cooling transition with a linewidth of $\Delta\omega = \omega_0\sqrt{(8kT\log2/mc^2)}$ in the range of $\Delta\omega \sim (2\pi)$ 1.5GHz. The repumping laser at 866nm driving the transition D$_{3/2}$ $\leftrightarrow$ P$_{1/2}$ to prevent ion loss is tuned on resonance. The power of 30μW for 397nm is focused to a spot size of 30μm, well below the saturation limit to avoid power broadening. The 866nm laser is focused to a spot size of 30μm with a power of 0.5mW.

The Doppler cooling limit of a few mK is achieved corresponding to half of the natural transition linewidth. For the fluorescence detection with the Doppler cooling laser the transition is saturated using a power of $300\mu W$ for maximum photon scattering, for Doppler cooling to the Lamb-Dicke regime the laser is attenuated respectively.

The quantized fluorescence signal from the trapped ions during photoionization loading (Fig. 7.1) shows the single ion resolution of the binned fluorescence signal of an ion cloud. The signal-to-noise ratio of ~ 25 (Fig. 7.1a) affirms the suppression of scattering light from background, which is necessary for single ion pictures (Fig. 7.1b). A single ion leads to a photon count rate of about 14kHz. After the phase transition of the ion cloud to a crystalline structure of the linear ion string, the Doppler cooling laser is tuned closer to the atomic resonance, because the linewidth of the excitation is narrowed. The radial ponderomotive confinement of the trap is always stronger than in the axial direction, so the linear ion string is oriented parallel to the control electrodes. The successive loading at a constant trap potential decreases the spacing of the equilibrium positions. The inner spacing of the two-ion crystal is about $7\mu m$.

Mixed species ion crystals are trapped rarely because of the photonionization loading technique. The impurity ions like CaH^+ oder $CaOH^+$ are identified by their non-fluorescing behavior with light at 397nm, but are trapped and sympathetically cooled by the remaining $^{40}Ca^+$ ions. The mass of the impurity ions can be determined exactly by comparison of the motional frequencies with $^{40}Ca^+$. The higher mass m leads to trap parameters of lower q and results in trajectories of a larger phase space area, so the occupation of the equilibrium positions is time-dependent because of heating effects.

7.2 Quantum jump spectroscopy

The metastable quadrupole transition $S_{1/2} \leftrightarrow D_{5/2}$ allows spectroscopy based on electron shelving. The ion is excited from the ground state $S_{1/2}$ with the 729nm laser by a weak transition to the $D_{5/2}$ state. The quantum jumps [Nag86, Sau86, Ber86] exemplify the long lifetime of the metastable level (Fig. 7.2a). Hence the $D_{5/2}$ state is depleted by the laser at 854nm. The decay of the $P_{3/2}$ state is shortened to the ground state $S_{1/2}$ by the strong dipole transition. The carrier and the sideband transitions are resolved dependently on the spectral linewidth of the 729nm laser. For a spectroscopy measurement the laser at 729nm is detuned from the carrier transition using an AOM to detect the secular and axial motional frequencies of a single ion confined in the trapping potential.

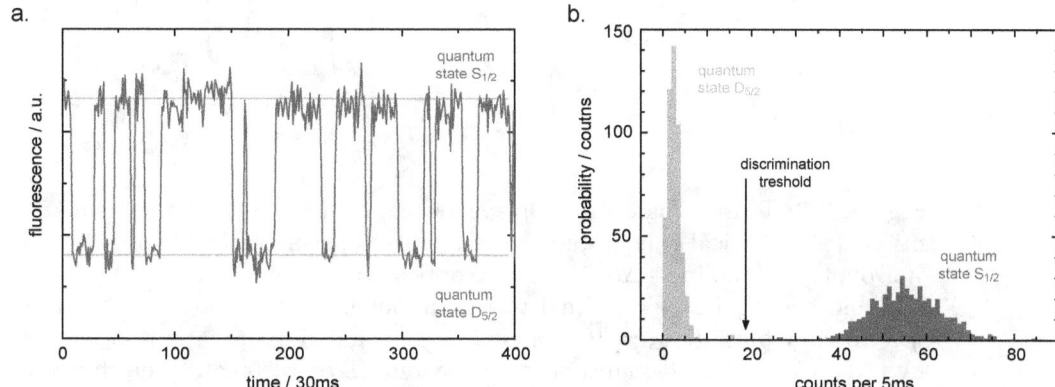

Figure 7.2: (a) The time-dependent resonance fluorescence signal near 397nm shows quantum jumps due to the excitation with the 729nm laser. The visibility of the quantum jumps is determined by the suppression of the background noise at the dark state. (b) Histogram of the fluorescence near 397nm for a single ion. The separated Poissionian distributions of the fluorescent $S_{1/2}$ state and the state $D_{5/2}$ allow an efficient discrimination. Until the number of photon counts of a single measurement exceeds the discrimination limit, the ion is assigned to the state $S_{1/2}$. The histogram is plotted using 1000 identical measurements at a count time of 5ms.

Because of the low scattering rate of the metastable state $D_{5/2}$ the occupation is determined by electron shelving with a statistical binary measurement of the fluorescence using the photomultiplier. In the $S_{1/2}$ ground state the ion is fluorescing by driving the transition for Doppler cooling at 397nm, but the fluorescence is absent if the ion is excited to the metastable $D_{5/2}$ state. The $D_{5/2}$ state occupation is measured with a fixed number of iterations as a mean value of all binary results of the photomultiplier detection cycles. For an efficient statistical detection a calibration measurement results in a treshold value for the future discrimination of the photomultiplier counts for direct state detection. The histogram of the calibration measurement (Fig. 7.2b) shows the background counts caused by straylight of the Doppler cooling laser at 397nm without ion fluorescence. At a higher fluorescence level the $S_{1/2}$ ground state occupation is detected. The distribution of the peaks follows Poissonian statistics. Dependent on the noise level of light at 397nm and the efficiency of the optical detection system both peaks are separated clearly, representing a single trapped ion.[1]

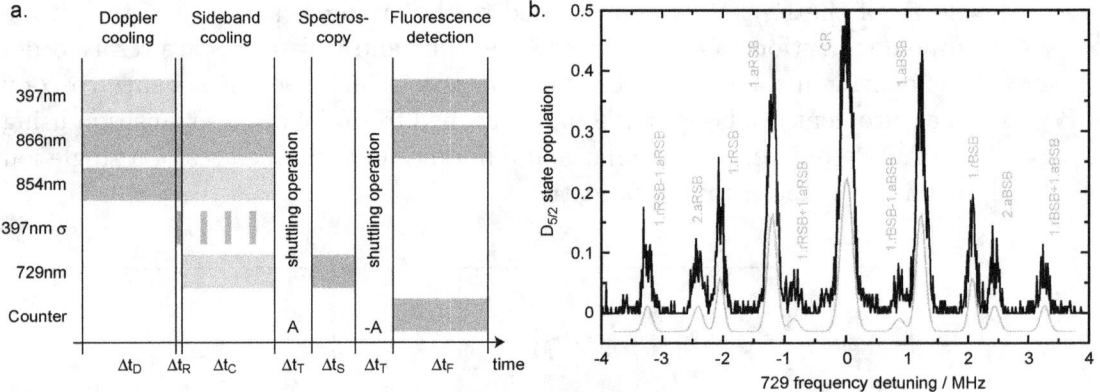

Figure 7.3: Pulsed quantum jump spectroscopy on the $S_{1/2} \leftrightarrow D_{5/2}$ transition: (a) A typical time sequence for a single measurement consists out of Doppler precooling ($\Delta t_D = 5$ms), optical pumping ($\Delta t_R = 100\mu$s), sideband cooling ($\Delta t_C = 6.5$ms) to the motional ground state, a 729nm spectroscopy pulse ($\Delta t_S = 100\mu$s) enclosed by shuttling operations A and -A ($\Delta t_T = 5$ms) and quantum state readout ($\Delta t_F = 5$ms). The different gray scales illustrate a tuned laser power or frequency. (b) The sideband spectrum is achieved by a 729nm laser detuning during Δt_S. The carrier transition (CR) and the first radial (1.rRSB, 1.rBSB) and axial sidebands (1.aRSB, 1.aBSB) are shown. The eqidistantly spaced second axial sidebands (2.aRSB, 2.aBSB) and the components 1.rRSB$\pm$1.aRSB and 1.rBSB$\pm$1.aBSB are resolved clearly. For the data shown here, a 729nm laser linewidth of $\Delta\nu = 200$kHz was used. The fit (shifted and downsized for clarity) calculates the axial and radial trap frequencies to $\omega_{ax} = (2\pi)\,1.2$MHz and $\omega_{ax} = (2\pi)\,2.0$MHz.

Based on the discrimination treshold for both quantum states $|0\rangle = |S_{1/2}\rangle$ and $|1\rangle = |D_{5/2}\rangle$ the population of the $D_{5/2}$ state is measured based on a variable detuning of the 729nm laser. For each detuning several identical time sequences (Fig. 7.3a) are averaged to a mean value μ representing the $|D_{5/2}\rangle$ state population. The statistical error $\sigma_{\bar\mu} = \sigma_\mu/\sqrt{n}$ for each detuning is given by the standard deviation $\sigma = \sqrt{\mu}$ and the number of iterations n of the pulse sequence.

---

[1]The histogram shows three peaks for two trapped ions. The peak of the background level and the peak for simultaneous fluorescence of both ions are located next to the center peak for the fluorescence of one of the two ions.

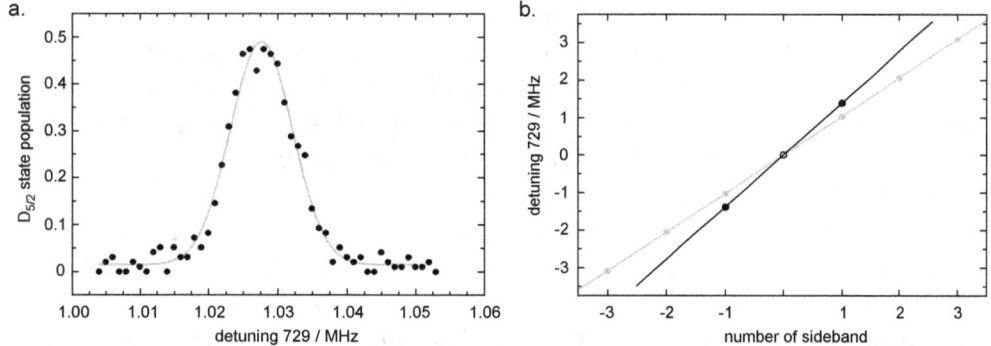

Figure 7.4: Characterization of trap potentials using quantum jump spectroscopy: (a) The upper axial motional sideband is measured at a frequency of 1.0289MHz, each data point represents n = 150 experiments. (b) The radial and axial trap frequencies are calculated with high accuracy based on the position measurements of different sidebands relative to the carrier (open circle). For a specified voltage configuration the radial trap frequency results in $\omega_{\mathrm{rad}} = 1.3860(4)$MHz (black), the axial trap frequency is evaluated to $\omega_{\mathrm{ax}} = 1.0291(3)$MHz (orange).

In the sideband spectrum measurement the spectroscopy is realized on the quadrupole transition $S_{1/2} \leftrightarrow D_{5/2}$ exclusively. With a lifetime of 1.2s, the spectroscopic resolution is limited by the laser pulse duration, the Rabi frequency during the excitation and the frequency stability of the laser source. The harmonic motion of the ion in the trap can be investigated spectroscopically at the level of single vibrational quanta. The electronic and vibrational states are manipulated coherently by the time sequence for pulsed spectroscopy (Fig. 7.3a). First, each time sequence (Fig. 7.3b) starts with the Doppler precooling on the dipole transition $S_{1/2} \leftrightarrow P_{1/2}$. The laser near 397nm is red-detuned near the half of the maximum fluorescence rate. This corresponds to a setting of about $\Gamma/2$, where $\Gamma = (2\pi)\,22.3$MHz is the natural linewidth of the dipole transition. The beam is attenuated in order to avoid saturation. The laser frequency near 866nm is tuned for maximum fluorescence. Additionally, resonant laser light depopulates the metastable $D_{5/2}$ level. Typically, Doppler precooling is applied for $\Delta t_D = 5$ms. A short optical pump pulse of the circular polarized 397nm laser ensure that the population is located in the Zeeman sublevel $|S_{1/2}, m = 1/2\rangle$ of the ground state, which is the starting point for the sideband cooling on the first motional red axial sideband.[2]

The 729nm laser is tuned to the red secular sideband of the $|S_{1/2}, m = +1/2\rangle \leftrightarrow |D_{5/2}, m = +5/2\rangle$ transition. The $D_{5/2}$ state is quenched by resonant laser light near 854nm to the $P_{3/2}$ level, which quickly decays to the ground state and closes the cooling cycle. Short pulses of optical pum-

---

[2]All measurements of the coherent single ion dynamics except for the determination of the heating rate are realized without sideband cooling.

ping are inserted and also conclude the sideband cooling. The optional ion shuttling algorithm A is applied to the control electrodes of the trap, then the 729nm spectroscopy pulse with the actual frequency detuning occur and the ion is shuttled back using the time-reversed algorithm -A. The quantum state readout is realized by photon scattering with the 397nm laser at maximum power to obtain a maximal count rate.

The 729nm laser stabilized to the high-finesse cavity resulting in a spectral linewidth of $\Delta\nu < 100$Hz allows a complete trap characterization using the sideband spectrum obtained with quantum jump spectroscopy (Fig. 7.4a). The positions of the equidistantly spaced radial and axial sidebands are measured and the the trap frequency are calculated with a uncertainty on the order of 10^{-6} (Fig. 7.4b).

7.2.1 Micromotion compensation

Field inhomogenities and patch charges on insulating areas lead to a displacement of the ion from the rf node of the electric field. Then the amplitude of the micromotion is increased significantly, the oscillation frequency is equal to the trap drive frequency of $\Omega/2\pi$. Consequently, the Doppler cooling and the fluorescence detection will be affected, because the enlarged amplitude results in a broader cooling transition. However, the micromotion is compensated by applying a balanced voltage difference on the particular segments of an electrode pair such that the ion is moved into the rf node of the trapping field (Fig. 7.5a). The numerical simulation of the electric potential for the micromotion compensation shows the equipotential lines from -0.7V to 0.7V with a regular increment of 0.05V. The control electrodes are biased with opposite sign for an nearly linear displacement parallel to the e_1-axis.

Various methods of measuring the micromotion exists [Ber98b] - three techniques were tested for the micromotion compensation of the microtrap. For fast micromotion compensation the single ions fluorescence is monitored by scanning manually the red side of the Doppler cooling transition. Simultaneously the fluorescence is enhanced varying the compensation voltage to narrow the cooling transition. For the experiments the Doppler cooling laser at 397nm is red-detuned to half of the linewidth using a static compensation voltage. A more quantitative approach of detecting the compensation voltage with the 397nm laser is achieved by a correlation measurement of the ions fluorescence rate with the phase of the rf trap drive. This technique deals with the dependency of the photon scattering to the frequency detuning of the cooling transition. Tuned to the inflection point of the transition, the largest gradient of the fluorescence curve will maximize the change of fluorescence induced by Doppler shifts of the 397nm laser at the moving ion. Because of the trap drive frequency of the micromotion the fluorescence depends on the rf phase. A time interval counter SR620[3] detects the fluorescence versus the rf phase. If a flatten line is detected, the ion is micromotion compensated perfectly. It was proven experimentally that the correlation measurement technique is equal in accuracy to the scheme for a fast compensation of micromotion, but has a longer detection time.

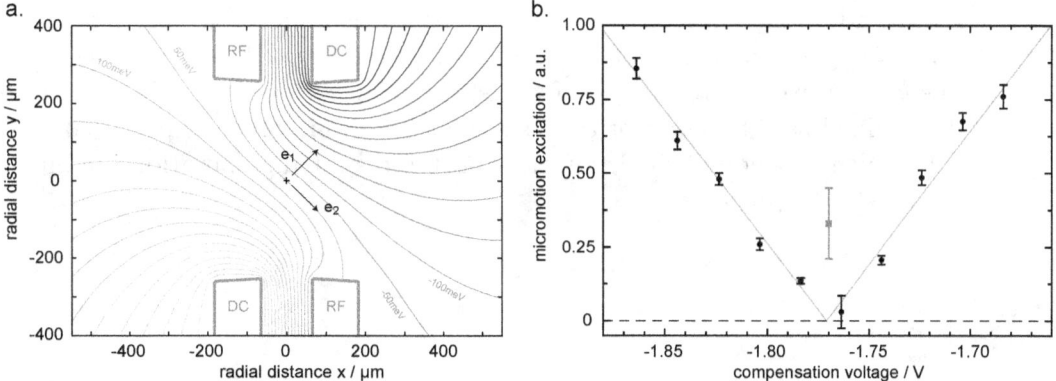

Figure 7.5: Compensation of micromotional sidebands: (a) The radial cross section of the electric potential for micromotion compensation is plotted at a compensation voltage of 1.5V. The dc electrodes are supplied with ±0.75V. The micromotion is compensated in the e_1-direction. (b) The amplitude of the micromotional sideband is measured for different compensation voltages. For comparison, the carrier excitation with a 1/10-reduced laser power is indicated by the single point.

The measurement of the micromotion using quantum jump spectroscopy is the method of highest accuracy: The detection of the first blue resolved sideband of the micromotion is achieved at the carrier frequency for the 729nm laser shifted by the trap frequency $\Omega/2\pi$. The excitation of the sideband is plotted against the compensation voltage and compared to the carrier excitation (Fig. 7.5b). The Rabi frequency Ω_1 on the micromotion sideband and that on the carrier Ω_0 holds $\Omega_1/\Omega_0 = J_1(\beta)/J_0(\beta) \approx \beta/2$ for the modulation index $\beta \ll 1$. As the excitation to the $D_{5/2}$ state is proportional to Ω^2 such for low saturation, the micromotion of the ions is measured directly. The minimum of excitation is close to -1.77V. At this minimum, thus for optimum compensation voltage, the ratio of excitation strength is reduced to zero with an error of ± 0.03. This measured value corresponds to the ratio of the squared Bessel functions $J_1^2(\beta)/J_0^2(\beta)$ with a modulation index β of 0.0 ± 0.17, which is due to a residual micromotion oscillation amplitude $\beta_{\min}(\lambda/2) = (0.0 \pm 0.17)(729\text{nm}/2)$ of 0.0 ± 130nm. The optimal compensation voltage changes by less than 0.5% from day to day in a time interval of one year.

In conclusion, the micromotion can be nulled by a properly compensation voltage and, due to the shielding effect of the gold coated finger shaped

---

[3]Standford Research Systems, Sunnyvale, USA

electrodes, stray electric fields from isolating parts of the microtrap are not disturbing the trapped ion by large and fluctuating contributions. Especially for the segmented microtrap the non-monolithic fabrication process leads to deviations from the ideal trap electrode geometry. Geometrical deviations achieve localized variations of the quadrupole trapping potential, which will induce additional micromotion on the ion. The compensation voltage for minimizing the ions micromotion depends in the end from the axial trap position. With the technique of transport spectroscopy the voltage can be determined as a function of the axial trap position, which is important for the implementation of single ion transport and splitting operations on linear ion crystals in future.

7.3 Coherent single ion dynamics

The sideband resolved measurement for the compensation of micromotion shows the application of quantum jump spectroscopy. The carrier, axial and radial motional sidebands of the Doppler cooled single ion are identified and the micromotion sideband is minimized with a specific compensation voltage. The coherent ion-light interaction leads to Rabi oscillations between the $S_{1/2}$ ground state and the excited metastable $D_{5/2}$ state by varying the excitation pulse length of the 729nm laser instead of the frequency detuninig. If the internal states $S_{1/2}$ and $D_{5/2}$ of the single ions are used to store qubit information, a π-pulse will flip the qubit between the two logic states $|0\rangle$ and $|1\rangle$. The $\pi/2$-pulse will generate a $(|0\rangle + |1\rangle)/\sqrt{2}$ state.

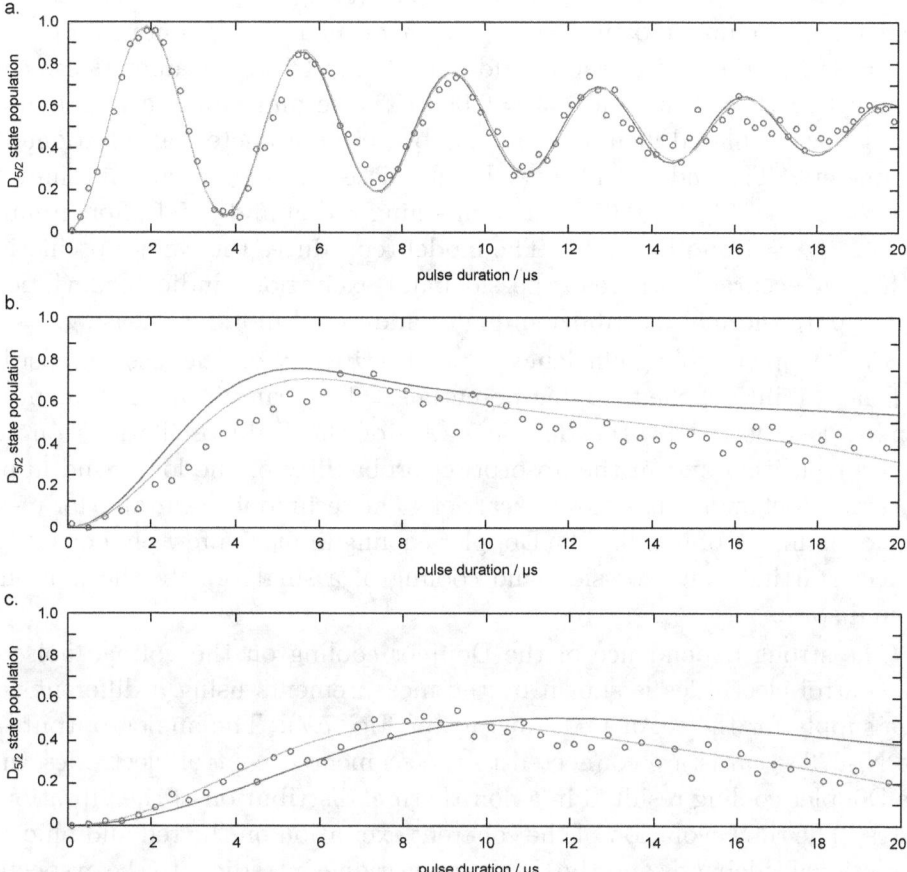

Figure 7.6: Coherent dynamics of a Doppler-cooled single ion: (a) Rabi oscillations up to 97% excitation at for a π-pulse at $1.88\mu s$ on the carrier transition and the measurement of the corresponding upper axial (b) and radial sideband (c). Each data point is representing 250 experiments. The numerical simulations are calculated with (orange) and without (gray) the Lamb-Dicke values as additional fit parameters (see text).

The coherent dynamic measurements are performed by the 729nm laser for coherent interaction with the internal states of the single ion and the following quantum state readout (Fig. 7.6). The pulsed 729nm excitation has a rectangular shape which pulse length is varied up to $30\mu s$. The $D_{5/2}$ state population shows Rabi oscillations on the carrier transition (Fig. 7.6a) with the axial (Fig. 7.6b) and radial (Fig. 7.6c) blue motional sidebands. The p-pulse on the carrier transition shows a efficiency of 97% at $1.88\mu s$, corresponding to a Rabi frequency of $\Omega_0 = (2\pi)\,923.1\mathrm{kHz}$. Each data point represents 250 experiments at a 729nm laser power of 60mW. The dephasing of the oscillation (a) resulting in a decay of contrast is due to an incoherent superposition of Rabi oscillations for each thermally occupied Fock state $\Omega_{n,n} \propto 1 - \eta^2 n$. The first numerical simulation (gray) is based on a thermal distribution p_n with $P(t) = \sum p_n(\bar{n})\sin^2(\Omega_{n,n}(t))$. The fit uses a mean axial phonon number of $\bar{n}_{ax} = 23$ and a mean radial phonon number of $\bar{n}_{rad} = 8$. The theoretical radial and axial Lamb-Dicke parameters are constant with $\eta_{rad} = 0.068$ and $\eta_{ax} = 0.066$. The second numerical simulation (orange) is calculated including the Lamb-Dicke parameters as additional fit parameters. The radial and axial Lamb-Dicke parameters are obtained to $\eta_{rad} = 0.066$ and $\eta_{ax} = 0.074$ with remaining radial and axial phonon numbers of $\bar{n}_{rad} = 7$ and $\bar{n}_{ax} = 14$. The model reproduces the carrier oscillation with high accuracy, whereas the sideband excitations indicate deviations from a pure thermal distribution due to additional motional heating.

The Doppler cooling efficiency depends critically on the laser intensities and the stability of the trapping potentials. The mean vibrational quantum number is reflected by the damping rate of the Rabi oscillations on the carrier transition and in the absorption probability of the lower and upper motional sidebands related to the carrier. The technique is suitable for mean phonon numbers of $\bar{n} < 5$. The Doppler cooling limit of a few phonons is the perfect starting point for sideband cooling of a single ion to the motional ground state.

The strong dependence of the Doppler cooling on the voltage noise of the control electrodes is shown by the measurements using a different voltage supply for the segmented microtrap (Fig. 7.7). The analog outputs of the NI-6733[4] cards are connected to the segmented control electrodes, and the Doppler cooling resulted in a non-thermal distribution of the vibrational states. The time evolution of the coherent excitation on the red and blue axial motional sidebands and the carrier transition contradicts to the exspected coherent evolution (Fig. 7.6). The imperfect Doppler cooling by the voltage noise on the control electrodes results in a strongly damped oscillation on the carrier transition (Fig. 7.7), while the red and blue axial motional sidebands show weakly damped Rabi oscillations. The $D_{5/2}$ state population is measured by varying the power of the 729nm laser at a fixed pulse length

---

[4]National Instruments, Texas, USA

of 20μs. Experiments with different pulse length at constant intensity of light at 729nm show the same effect. There, the π-pulse on the upper axial motional sideband is achieved at 20μs (Fig. 7.7b). The average (orange) illustrates the experimental data. The high oscillation strength of the first axial sidebands compared to the damped oscillation on the carrier indicates that the conditions for the Lamb-Dicke regime are not fulfilled (Fig. 7.7a). The mean phonon number is estimated to a couple of hundred quanta. The imperfect Doppler cooling shows an example of coherent dynamics on a single ion outside the Lamb-Dicke regime (Fig. 7.7). In the Lamb-Dicke regime the coupling strength of the carrier is dominating (Fig. 7.7a), resulting in strong Rabi oscillations on the carrier and rapidly damped oscillations on the sidebands (Fig. 7.6).

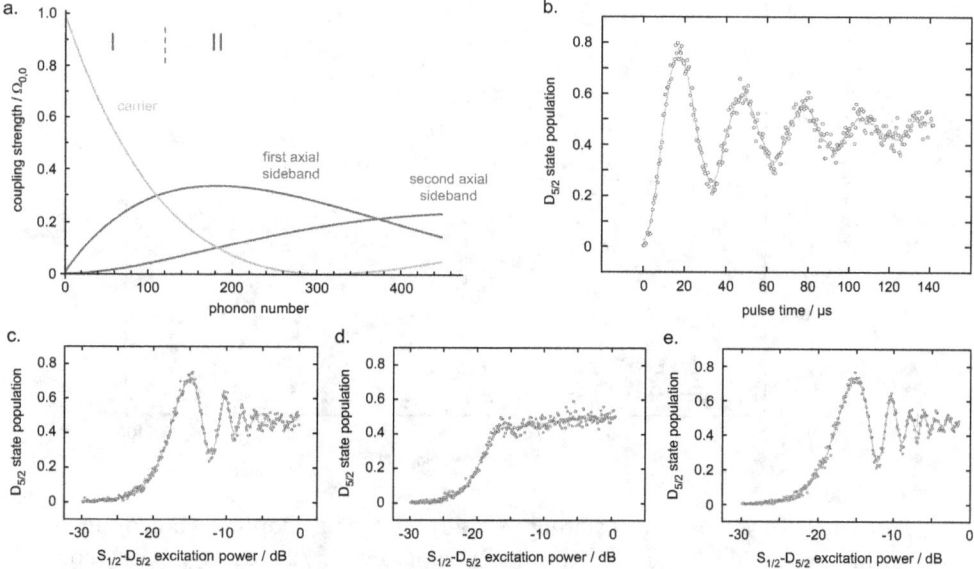

Figure 7.7: Coherent dynamics of imperfect Doppler cooling of a single ion: (a) The coupling strength shows the Lamb-Dicke regime (I), the sidebands are elevated outside of the regime (II). The upper axial motional sideband shows Rabi oscillations (b), on the carrier are their not existent (d). The excitation strength of the lower axial sideband (c) is approximately the same compared to the upper sideband (e) (cp. [Zzy]).

The coherence of the atomic system is proven by a Ramsey resonance measurement. The dephasing of the Rabi oscillation on the carrier shows the influence of decoherence, but a Ramsey experiment allows to quantify it precisely (Fig. 7.8): A pulse pattern of two $\pi/2$-pulses of defined duration and spaced by a constant pulse separation time is used for coherent excitation at different frequency detunings around the carrier transition. This pump and probe-type experimental technique is more advantageous than the measurement of Rabi oscillations on the carrier of identical time. The

Ramsey experiment is not influenced by systematic frequency drifts, i.e. of the 729nm laser or the magnetic field, because of the short excitation pulses combined with a long delay time compared to the measurement of Rabi oscillations by a permanent excitation of identical length. The Ramsey fringe pattern is characterized by the $\pi/2$-pulse time and respectively the corresponding Rabi frequency and the delay time between the two pulses. The envelope of the Ramsey pattern is determined by the properties of the $\pi/2$-pulses, the contrast of the fringe pattern is based on the pulse separation time. Within the pulse spacing the system evolves freely interacting with various decoherence mechanisms.

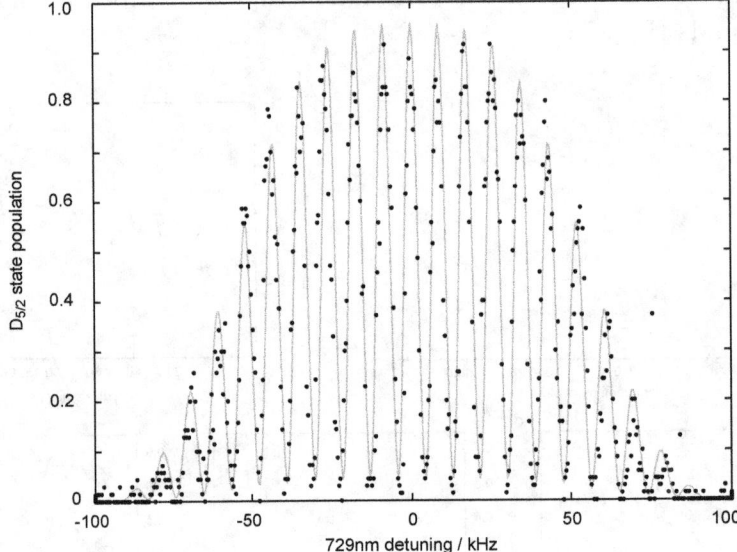

Figure 7.8: Ramsey interference experiment of a Doppler-cooled single ion in the loading region of the microtrap: Two $\pi/2$-carrier pulses separated by a delay time Δt are driven by the 729nm laser with a variable frequency detuning. The pulse duration is $t_p = 10\mu s$, the pulse separation time $\Delta t = 94\mu s$. The theoretical curve is calculated numerically from the time evolution of the density matrix with a purely transversal decay rate.

The Ramsey experiment of a single ion after Doppler cooling allows the determination of the decay rate. The numerical optimization yields a decay of $\gamma = (2\pi)\,0.8\text{kHz}$ with a Rabi frequency of $\Omega_0 = 10.9\text{kHz}$ and is in good agreement with the experimental data (Fig. 7.8). The major difference of the theoretical curve from the experimental data is caused by the short-time frequency drift of the 729nm laser. The contrast decay of the Ramsey fringe pattern is improved significantly by triggering the pulse sequence to the phase of the 50Hz power line (Fig. 7.9). A standard Schmitt-Trigger electronic circuit is used to start each experiment at the identical phase. In general the duration of the pulse sequence with quantum state readout included is less than 20ms, each pulse pattern will be executed in adjacent periods of the power line. The contrast of the Ramsey fringes is deduced from the oscillations near zero detuning (Fig. 7.9a) and determined for different pulse separation times (Fig. 7.9b). The Ramsey pulse sequence for this experiments is characterized by a $\pi/2$-carrier pulse of $10\mu s$, a delay time of $94\mu s$ and a final $\pi/2$-carrier pulse. The influence of the frequency noise at 50Hz on the Ramsey contrast by ambient magnetic field fluctuations is shown with an active power line trigger (circle) and without triggering (square). The trigger improves the pulse separation time at a Ramsey contrast of 0.4 from $160\mu s$ to $700\mu s$. The enhancement of the contrast is independent from the initial phase of the power line, which is proven experimentally.

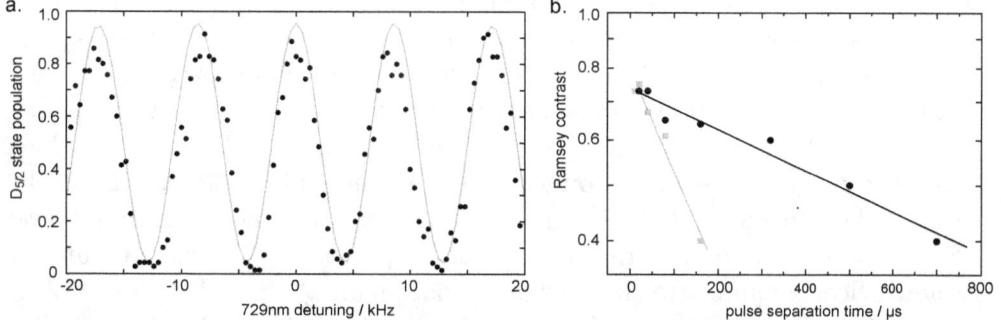

Figure 7.9: Measurement of the Ramsey contrast decay: (a) The Ramsey contrast is defined as the difference of the maximal and minimal $D_{5/2}$ state population at the carrier transition. The experimental data and the theoretical curve is part out of Fig. 7.8. (b) The Ramsey contrast decays exponentially with the increase of the pulse delay (orange). The contrast is preserved significantly using a power line trigger to minimize the decay by ambient magnetic field fluctuations (gray).

The power line trigger is prerequisite for the investigation of decoherence effects on the atomic qubit - the statistical nature of the quantum state readout requires various identical measurements, so the initial conditions are optimized using a phase-dependent experimental sequence with respect to the ambient magnetic field fluctuations. An actively stabilized magnetic field compensation or - in the end - a passive mu-metal shielding with a high magnetic permeability would enhance the decoherence of the qubit.

In general the relevant coherence-limiting processes are fluctuations of the 729nm laser intensity and phase as well as fluctuations of the static magnetic field. The noise induced by the variation of the magnetic field shifts the frequency of the quadrupole transition [Zzz], fluctuations of the laser power spreads out the Rabi frequency and results in a fast dephasing of the Rabi oscillations. Especially in segmented microtraps the stability and the noise of the different control voltages induces vibrational heating on the trapped ions. A stable voltage supply for the control electrodes is crucial for confining the trapped ions within the Lamb-Dicke regime.

7.3.1   Sideband cooling

The application of quantum algorithms requires often ions cooled to the motional ground state. The temperature limit of Doppler precooling to the Lamb-Dicke regime is given by the natural linewidth of the dipole transition, which results in a mean phonon number of several residual quanta. The sideband cooling technique provides cooling of the Doppler-precooled ion to the motional ground state. The resolution of the sideband measurement of single quanta allows the determination of the trap heating rate.

Starting with Doppler precooling to the Lamb-Dicke regime, axial motional sideband cooling on the narrow $S_{1/2} \leftrightarrow D_{5/2}$ transition is demonstrated. The cooling laser at 729nm is tuned to the red axial sideband of the transition to excite single ions from the $|S_{1/2}, m = 1/2\rangle$ state to the metastable $|D_{5/2}, m = 5/2\rangle$ state. The effective width of this cooling transition is increased by applying resonant laser light near 854nm. By quenching with the 854nm laser, the $D_{5/2}$ state is mixed with the $P_{3/2}$ state, which decays rapidly to the $S_{1/2}$ ground state. At the end of the cooling cycle, a σ^+-polarized pulse of the 397nm laser depopulates the Zeeman sublevel $|S_{1/2}, m = -1/2\rangle$ to prepare the ion in the $|S_{1/2}, m = 1/2\rangle$ ground state for a new cylce to annihilate the next motional quanta.

Here, the laser cooling of the trapped ions is investigated by the semiclassical theory [Ste86], all laser cooling parameters are determined experimentally and the cooling result is compared with the theorectial expectation. The theoretical limit of sideband cooling is given by the ratio of the laser cooling Γ_{cool} and the heating rates $\Gamma_{heat} = \Gamma_{laser} + \Gamma_{trap}$, yielding to an average steady-state phonon number of $\bar{n} = \Gamma_{heat}/(\Gamma_{cool} - \Gamma_{heat})$. Heating by laser processes expressed by Γ_{laser} is due to either an off-resonant excitation on the carrier transition with subsequent decay on the blue sideband or an off-resonant blue sideband excitation followed by a decay on the carrier [Ste86]. A calculation of the detailed balance of phonon states leads to

$$\bar{n} = \left(\frac{\eta_{spont}^2}{\eta_{729}^2} + \frac{1}{4} \right) \frac{\gamma_{eff}^2}{4\omega} \, , \tag{7.1}$$

if trap heating Γ_{trap} is excluded. The parameter η_{spont} takes the recoil into account if the ion decays spontaneously from the $P_{3/2}$ level to the ground state $S_{1/2}$. For a motional axial trap frequency of $\omega_{ax} = (2\pi)\,1.1\mathrm{MHz}$ it is calculated to $\eta_{spont} = |\tilde{k}_{395}| \cdot \sqrt{\hbar/2m\omega_{ax}} = 0.17$. Due to the laser recoil on the $S_{1/2}$ to $D_{5/2}$ excitation, the Lamb-Dicke factor results in $\eta_{729} = 0.065$. In equation (7.1) only the cooling rate, not the cooling limit, depends on the intensity of the 729nm laser. The parameter γ_{eff} denotes the effective linewidth from quenching the $D_{5/2}$ state to the $P_{3/2}$ level and is adjusted by the laser power at 854nm.

The average phonon number $\bar{n}$ increases with the intensity of the 854nm laser, thus one has a tradeoff between cooling rate and minimum temperature. Even for $\gamma_{\text{eff}} \simeq 90\text{kHz}$, the equation (7.1) predicts an almost perfect ground state of vibration with only $\bar{n} \leq 0.01$. The situation is more complicated if trap heating Γ_{trap} is taken into account: The steady-state phonon number $\bar{n}$ results from the balance

$$\bar{n} = \frac{\Gamma_{\text{laser}} + \Gamma_{\text{trap}}}{\Gamma_{\text{cool}} - \Gamma_{\text{laser}} - \Gamma_{\text{trap}}} \simeq \frac{\Gamma_{\text{trap}}}{\Gamma_{\text{cool}} - \Gamma_{\text{trap}}} \,, \tag{7.2}$$

if the laser induced heating Γ_{laser} is small compared to the trap heating Γ_{trap} and cooling rate Γ_{cool}. We find that the thermal mean phonon number $\bar{n}$ becomes

$$\bar{n} = \frac{\Gamma_{\text{trap}}}{(\eta_{729}\Omega_0/\gamma_{\text{eff}})^2 \ \gamma_{\text{eff}} - \Gamma_{\text{trap}}} \,, \tag{7.3}$$

for the case where the net cooling rate $W = (\eta_{729}\Omega_0)^2/\gamma_{\text{eff}} - \Gamma_{\text{trap}}$ is positive. In contrast to the equation (7.1), the cooling limit now depends on the intensity of the laser near 729nm driving the red sideband of the quadrupole transition. We consider here the case where the sideband excitation is incoherent, with $\eta_{729}\Omega_0 \leq \gamma_{\text{eff}}$.

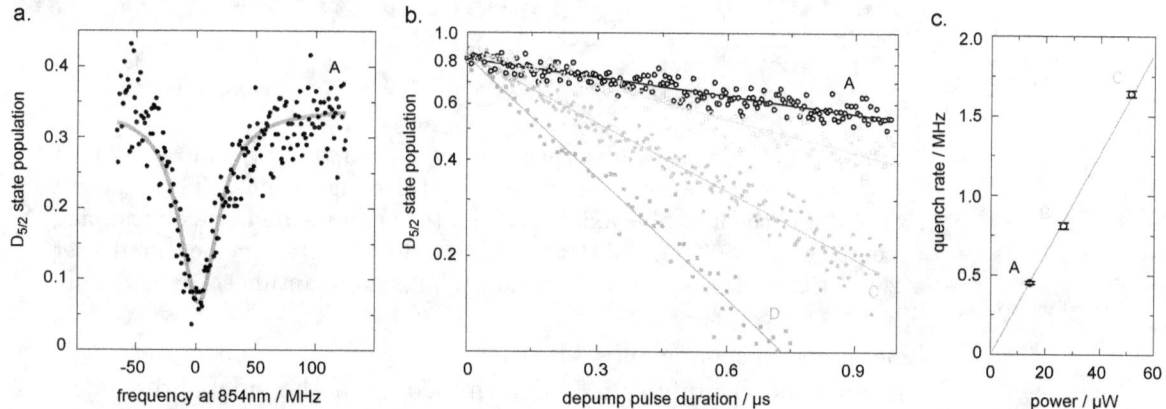

Figure 7.10: Depletion of the $D_{5/2}$ state by a 854nm laser pulse with $3\mu s$ duration: (a) The dip corresponds to the resonance line and allows tuning the laser to resonance. The data is modeled by a 37MHz power-broadened Lorentzian. (b) After a π-pulse on the carrier transition at 729nm, the depletion pulse length to the $P_{3/2}$ state is scanned. The exponential decay determines γ_{eff}, here plotted at four different 854nm laser powers. (c) The linear dependence on the quench rate γ_{eff} to the 854nm laser power yields to a linear constant of $31.6(5)\text{kHz}/\mu\text{W}$.

In order to compare the cooling theory to the experiment, γ_{eff}, Ω_{729} and Γ_{trap} are determined by independent measurements. The theoretical prediction in equation (7.3) is compared with the experimental outcome on $\bar{n}$. The quench rate γ_{eff} is obtained by controlled depletion of the metastable state (Fig. 7.10): After Doppler cooling and optical pumping, a 729nm laser pulse of $1\mu s$ duration is applied on the carrier transition to transfer the ion into the $D_{5/2}$ state. A pulse of the 854nm laser resonant to the $D_{5/2} \leftrightarrow P_{3/2}$ transition is then applied and finally the remaining $D_{5/2}$ population is detected. The exponential decay (Fig. 7.10b) determines the effective cooling width that is as expected a linear function of the laser power at 854nm (Fig. 7.10c). The Rabi frequency Ω_0 on the quadrupole transition is revealed from measurements of Rabi oscillations (Fig. 7.6a). With the maximum available 729nm laser power of 60mW a Rabi frequency of $\Omega_0 \simeq (2\pi)\,200\text{kHz}$ is achieved. Correspondingly, the maximum possible sideband excitation with this laser power is $\eta\Omega_0 = (2\pi)\,13\text{kHz}$.

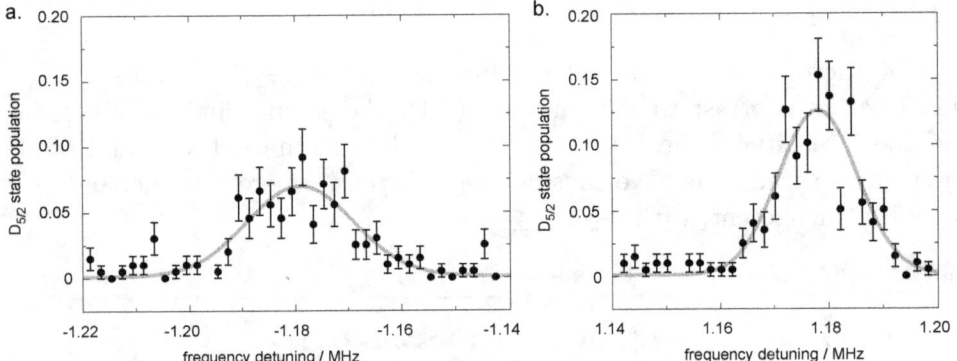

Figure 7.11: Sideband absorption spectrum on the $|S_{1/2}, m = 1/2\rangle \leftrightarrow |D_{5/2}, m = 5/2\rangle$ transition after sideband cooling of 5ms. The first red axial (a) and the first blue axial sideband (b) are measured at a dc trapping potential of $\omega_{\text{ax}} = (2\pi)1.18\text{MHz}$ frequency. Both spectra are measured after a $200\mu s$ time delay between sideband cooling and quantum state detection.

The sideband cooling allows further cooling of the Doppler-precooled ion. The mean phonon number $\bar{n}$ is determined from the sideband absorption spectrum: The sideband absorption S_r on the red axial sideband depends from the mean phonon number $\bar{n}$ as $S_r \propto \bar{n}$, the blue sideband absorption S_b like $S_b \propto \bar{n}+1$. At the motional ground state the red sideband disappears in the absorption spectrum, the blue sideband remains with small absorption strength. The vibrational energy of the ion corresponds to the intensity ratio of the lower and upper motional sideband absorptions with a phonon number of $\bar{n} = (S_r/S_b)/(1 - (S_r/S_b))$ [Mon95a]. The measurement (Fig. 7.11) shows sideband cooling to a mean phonon number of $\bar{n} = 1.2(3)$. The temperature of the ion $kT = \hbar\omega_{\text{ax}}/\ln(1 + 1/\bar{n})$ results to $T = 34(8)\mu K$ [Die89].

7.3.2 Heating rate measurements

The coherence of the experimental system is investigated using quantum jump spectroscopy after sideband cooling of the Doppler-precooled single ion. The single ion cooled to the axial motional ground state serves as a probe with unexcelled accuracy of single motional quanta. The detection of the lower and upper axial motional sideband results in the mean phonon number $\bar{n}$. Inserting a variable delay time between sideband cooling and quantum state readout, the mean phonon number $\bar{n}(\Delta t)$ depends on the delay time Δt, in which the external experimental system interacts with the ion. Assuming a thermal state, the motional decoherence of the ion is linear to the waiting time.

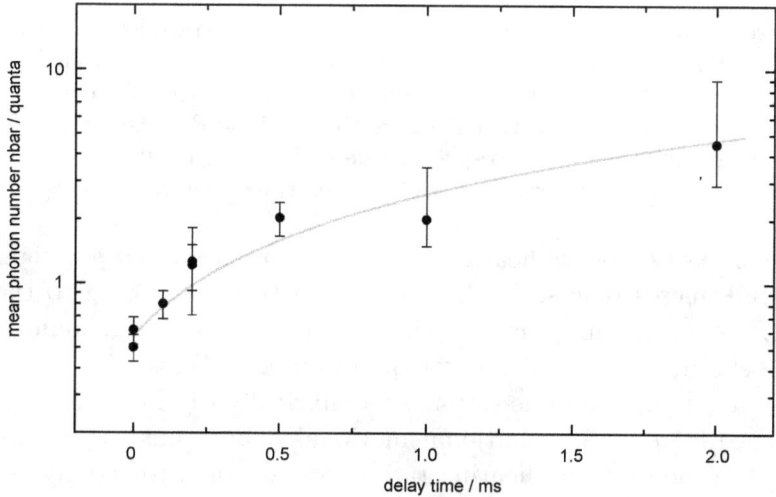

Figure 7.12: Heating rate measurement for the axial vibrational mode of a single ion at the loading region of the microtrap: the cumulative mean phonon number $\bar{n}$ is measured after a variable delay time Δt between sideband cooling and quantum state detection.

The heating rate is measured in the loading region of the segmented microtrap for a constant radial and axial confinement with a single ion (Fig. 7.12). The time duration of the sideband cooling has a fixed length of 5ms while the delay time changes. The ground state occupation represented by the mean phonon number $\bar{n}(0s)$ without delay between cooling and readout is $\bar{n} = 0.56(5)$ quanta. The axial trap heating rate is deduced from a linear fit to $\bar{n}/\Delta t = 2.1(3)$ quanta per millisecond at an axial trap frequency of $\omega_{ax} = (2\pi)\,1.18\text{MHz}$. As this trap heating rate is attributed to the permanent contamination of the trap electrodes due to the loading process from the thermal beam of neutral calcium [Zzz], we exspect a significant lower mean phonon number $\bar{n}$ in the processing region, which is spatially separated from the loading zone.

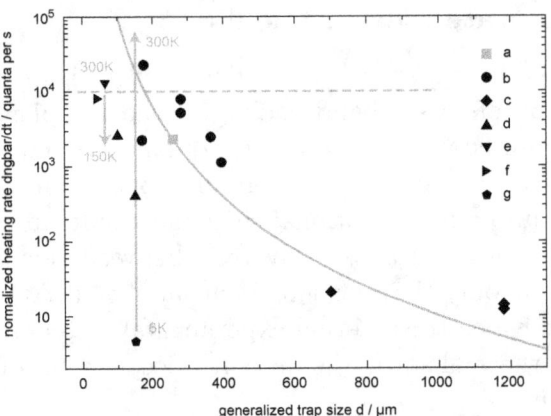

Figure 7.13: Normalized heating rates for various trap designs based on macro- and microfabrication techniques: The heating rate is scaled to the ion mass of $^{40}\text{Ca}^+$ and an axial mode frequency of 1MHz for comparison. The generalized trap size is defined as the minimal distance between the trapped ion and the nearest trap electrode surface and allows a classification of the trap scale. The details of A-G are described below.

The comparison of the heating rates $\bar{n}/\Delta t$ for various trap designs of the macro- and microscopic scale classify the ion traps regarding to the applicability for quantum information science. The trap heating should be lower than the execution time needed for quantum algorithms. A rough limit of $100\mu s$ for a gate operation is estimated empirically (dashed line), where the decoherence in terms of additional motional quanta has to be suppressed (Fig. 7.13). The motional heating rate is proportional to the inverse of the trap frequency ω and mass m of the trapped ions. For a comparison of different traps the calculation of a normalized heating rate $\bar{N}/\Delta t$ is required. Therefore the heating rate $\bar{n}/\Delta t$ is scaled with ω^{-1} and m^{-1}. It is proportional to the electric field noise density $S_E = 4\hbar m\omega/e^2 \cdot \bar{n}/\Delta t$ [Lab08b]. Even the electric field noise S_E is proportional to the net voltage noise S_V, which is an incoherent sum of all sources of Johnson noise [Des06].

Empirically the trap heating depends on the minimal distance d between the electrode surface and the ion with the scaling $\bar{N}/\Delta t \propto 1/d^4$ (Fig. 7.13). This characteristic curve is plotted for the normalized heating rate of the microtrap (A). Macroscopic ion traps show a significant low trap heating (B), but are not scalable [Tur00]. Improvements on the macroscopic trap designs (C) are located in the same regime [Roh01]. Simple microfabricated traps (D) [Des04] and traps consisting out of two needles (E) [Des06] are at the upper decoherence limit. The planar traps (F) are located near this limit [Sei06] or far beyond (G). Experiments with the needle trap design show the influence of cooling (E), the planar trap operation in a cryogenic environment at 6K decreases the trap heating by a factor of 10^5 compared to room temperature [Lab08a].

7.3.3 Robust adiabatic passage

The standard pulse sequence for quantum jump spectroscopy is characterized by time-dependent laser pulses of rectangular shape regarding the laser power. The spectroscopy pulse of the 729nm laser for coherent quantum state manipulation is controlled by a double pass AOM configuration. In standard configuration, an rf switch triggers the rf supply of the 75MHz AOM. To improve the robustness of the quantum state readout, the pulse is implemented with arbitrary amplitude and envelope (Fig. 7.14a). The amplitude and phase of the pulse is transferred directly to the light field.

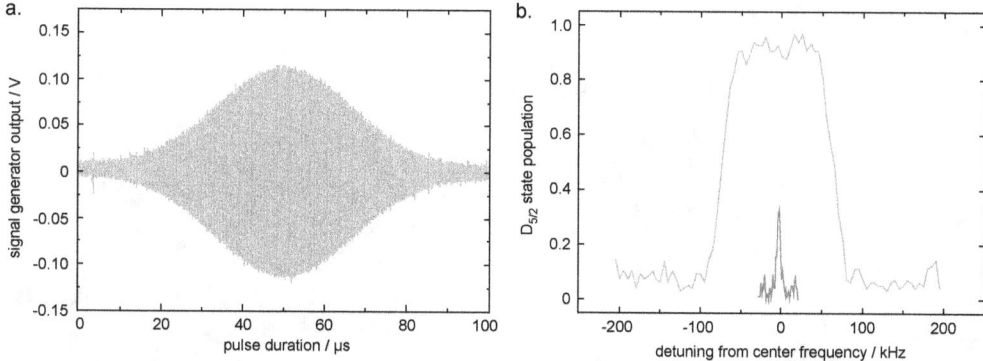

Figure 7.14: Coherent population transfer using a robust adiabatic passage (RAP) pulse: (a) The measured RAP transfer pulse is a Gaussian-shaped frequency chirp with a center frequency of 75MHz. The pulse length is 100μs, the frequency detuning 130kHz. (b) The $D_{5/2}$ state population of the carrier transition is shown for different detunings using a 100μs rectangular pulse (gray) or the RAP transfer pulse (orange).

The robust adiabatic passage (RAP) couples the levels $S_{1/2}$ and $D_{5/2}$ using a frequency-chirped 729nm laser pulse. The frequency chirp is linear in time and centered at the carrier frequency. The time duration Δt of the chirp is 100μs, the overall frequency detuning $\Delta \nu = 130$kHz. The time-dependent amplitude is proportional to an envelope $\exp(-(t - \Delta t/2)^2/4\sigma^2)$ of a Gaussian shape with $\sigma = \Delta t/6\sqrt{2}$ [Wun07]. The pulse is generated by an arbitrary waveform synthesizer VFG-150[5] with a 16-bit resolution and a timebase of 5ns.

The measurement (Fig. 7.14b) shows a population transfer with the RAP pulse compared to a normal rectangular pulse. The population transfer from the $S_{1/2}$ to the $D_{5/2}$ state is very efficient over a large detuning of 160kHz compared to 10kHz for the standard pulse. The RAP pulse provides a robust and reliable population transfer against variations and noise at experimental parameters and is combined with the frequency stabilized 729nm laser. The frequency shifts of the 729nm laser are levelled by the broad effi-

---

[5]Toptica AG, Gräfelfing, Germany

cient detuning range, so the linewidth on the order of 100Hz for the 729nm laser is sufficiant and the effort of an improvement of the laser stabilization is limited. Other techniques like composite pulses developed in NMR research have similar properties and can be transferred to ion trap experiments. The application of RAP pulses as well as CORPSE (compensation for off-resonance with a pulse sequence) [Tim08] or SCROFULOUS (short composite rotation for undoing length over and under shoot) [Tim06] composite pulses can improve the population transfer significantly.

7.4 Transport spectroscopy

The complete characterization of the multi-segmented microtrap requires an enhanced technique for the measurement of motional frequencies and coherent manipulation of the ion at different axial positions. Future experiments concerning scalable quantum information science will integrate shuttling operations in the pulse sequence of the quantum jump spectroscopy. For the first time the presented so-called transport spectroscopy is a prospective study for the hybridization of the experimental time sequence concerning transport and spectroscopy.

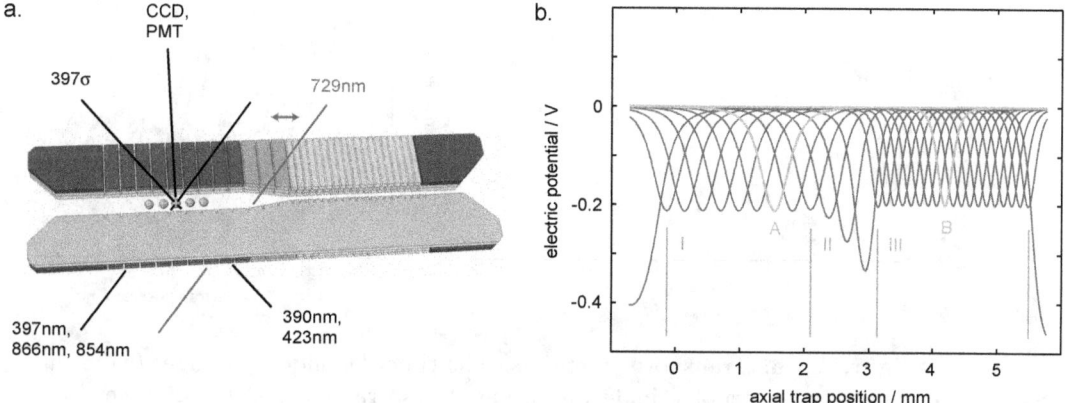

Figure 7.15: Transport spectroscopy: (a) Loading, Doppler cooling and quantum state readout are localized at the same electrode pair, the 729nm spectroscopy laser is movable to different axial positions. (b) The individual calculated electrode potentials for each electrode pair at −1V are prerequisite for simulating smooth ion transports. The storage zone (I), the transfer electrodes (II) and the processing zone (III) are controlled independently.

The photoionization loading, Doppler cooling and quantum state readout remain at the same fixed position in the center of the storage region. The 729nm laser for coherent state manipulation is moved along the axial trap direction to a specific electrode pair (Fig. 7.15a). The operation scheme starts with loading of a single ion and Doppler cooling at the specific storage segment. The single ion is shuttled with the control voltages to be addressed with the spatial separated 729nm laser. The 729nm laser imprints the motional frequencies at this dedicated point to the ions quantum state. Then the ion is shuttled back by the inverse transport sequence followed by the quantum state readout. The excitation with the 729nm laser allows a very precise measurement of all motional frequencies of the ion at a specific point in the trap, with respect to a fast and efficient characterization of the microtrap.

The calculation of the transport sequence requires accurate knowledge of the electric potential of each electrode segment (Fig. 7.15b). The electric potentials are calculated from a numerical simulation using boundary element methods. Bounded by the endcap electrodes, the storage region potentials (I) are more spatial extended (A) than the potentials (B) in the processing region (III). The three transfer electrode segments (II) will influence the ion especially towards the storage region.

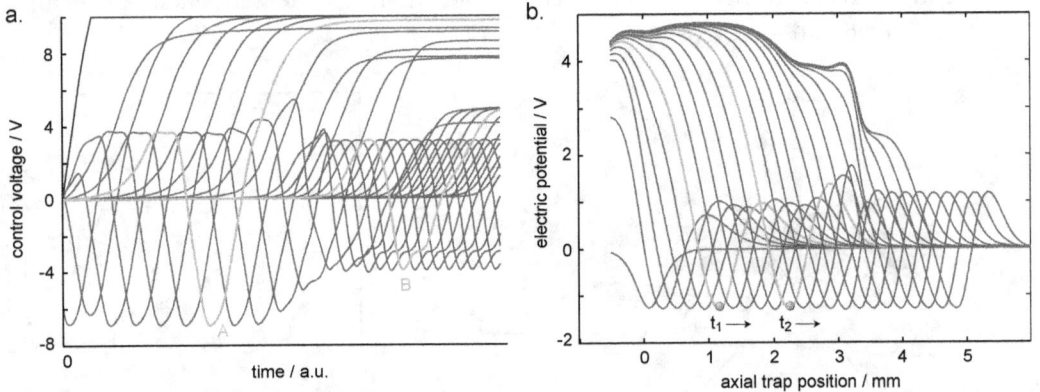

Figure 7.16: Transport potentials: The time-dependent voltages (a) for a smooth transport of a single ion through the storage, transfer and processing region are calculated. The voltage waveforms for each electrode pair are shown. The axial motional frequency of $\omega_{ax} = 1.13$MHz remains constant, snapshots of the cumulative potential (b) are plotted on a equidistant spaced timescale ($t_1 < t_2$).

The time-dependent electric voltages for each electrode pair are calculated with the constraint that the axial frequency $\dot{\omega}_{ax} = 0$ is preserved during the transport operations. The calculation scheme of the time-dependent voltages is based on a singular value decomposition [Gol89]: Assuming the intrinsic electric potential $\varphi_k(x)$ of the k-th electrode, which is obtained from the numerical simulation by applying +1V and ground the adjacent electrodes, the voltage $U_k(t)$ controls the strength along the axial trap direction x. The total electric potential $\Phi(x, t)$ of all electrodes interacting with the trapped ion results to $\Phi(x, t) = \sum_k \varphi_k(x) \cdot U_k(t)$. The axial trap direction x and the time t are discretized for the numerical calculation as $x \in [x_0, \ldots, x_m] \in \mathbb{R}^m$ and $t \in [t_0, \ldots, t_n] \in \mathbb{R}^n$. The electric potential $\varphi_k(x_m) \in \mathbb{R}^{m \times k}$ defined by the trap geometry is adapted by the voltage $U_k(t_n) \in \mathbb{R}^{k \times n}$ to $\Phi(x_m, t_n) = \sum_k \varphi_k(x_m) \cdot U_k(t_n)$ with $\Phi(x_m, t_n) \in \mathbb{R}^{m \times n}$. The accuracy of the calculation is ensured with $\Delta x = x_m - x_{m-1} \approx 1\mu m$ and $\Delta t = t_m - t_{m-1} \ll 2\pi/\omega_{ax}$. In a simplified calculation the time t is separated, so each time step of the voltage $u_k \in \mathbb{R}^k$ is calculated successively. The overall electric potential $\phi(x_m) \in \mathbb{R}^m$ at the axial direction results to

$$\begin{pmatrix} \phi(x_1) \\ \vdots \\ \phi(x_m) \end{pmatrix} = \begin{pmatrix} \varphi_1(x_1) & \cdots & \varphi_k(x_1) \\ \vdots & \ddots & \vdots \\ \varphi_1(x_m) & \cdots & \varphi_k(x_m) \end{pmatrix} \begin{pmatrix} u_1 \\ \vdots \\ u_k \end{pmatrix} \tag{7.4}$$

The shuttling of the ion with a constant axial frequency ω_{ax} is modeled by the manual generation of an overall starting potential $(\phi(x_1), \ldots, \phi(x_m))$ by an initial set of voltages. The calculation of the voltages $(u_1, \ldots, u_k)$ for a each subsequent time step t_n is based on the inverse of $\varphi(x) \in \mathbb{R}^{k \times m}$ with

$$u = \varphi^{-1}(x)\phi(x) \tag{7.5}$$

The inverse of the rectangular matrix $\varphi(x)$ is obtained using singular value decomposition as a generalized method for eigen-decomposition, which is defined typically for squared matrices only. The fundamental idea is to decompose $\varphi(x)$ in two squared orthonormal matrices $A \in \mathbb{R}^{k \times k}, B \in \mathbb{R}^{m \times m}$ and a diagonal matrix $X \in \mathbb{R}^{k \times m}$ containing the so-called singular values. The orthonormal matrices A, B consist out of the normalized eigenvectors of $\varphi(x)\varphi(x)^+$ and $\varphi(x)^+\varphi(x)$ respectively. The matrix X is the diagonal matrix of the squared eigenvalues of $\varphi(x)\varphi(x)^+$. Then the potential matrix $\varphi(x) = AXB^*$ can be inverted easily with $\varphi(x)^{-1} = BX^+A^*$. For an accurate fit of the static axial frequency ω_{ax} the overall potential $(\phi(x_1), \ldots, \phi(x_m))$ has to be reproduced by the voltages $(u_1, \ldots, u_k)$ in a finite region around the potential minimum. The axial frequency is determined by a sixth-order polynomial fit in the finite region of interest.

The influence of the electrode pairs far away from the ion to the electric potential at the ions position is limited. Therefore the solution of this ill-posed problem can cause large voltage fluctuations in time at the outer electrode pairs which are avoided using the method of Tikhonov regularization: The singular value decomposition is extended using the voltage range as an additional constraint. This generalized singular value decomposition avoids diverging terms by calculating the inverse of the diagonal matrix $X = (\delta_{km}x_{km})$ with an additional constant regularization matrix $\Gamma = (\delta_{km}\xi_{km})$. It is levelled adequately and inserted in the diagonalization $X^{-1} = (\delta_{km}(x_{km}^2 + \xi_{km}^2)/x_{km})^{-1}$ to optimize the solution to the limited voltage range. The mathematical basis is to minimize the combination of the residual and the regularization matrix $||\varphi(x)u - \phi(x)||^2 + ||\Gamma u||^2$. The solution for a specific timestep is found if the axial motional frequency is located within the accuracy limit and the voltage range for all control electrodes is preserved. The voltages u for the following timestep are determined based on a new given overall potential $\phi(x)$ with a moved axial potential minimum and the old solution as the initial guess. This solution method of the ill-posed problem maintains the smooth expansion of the time-dependent voltage waveform for each trap electrode.

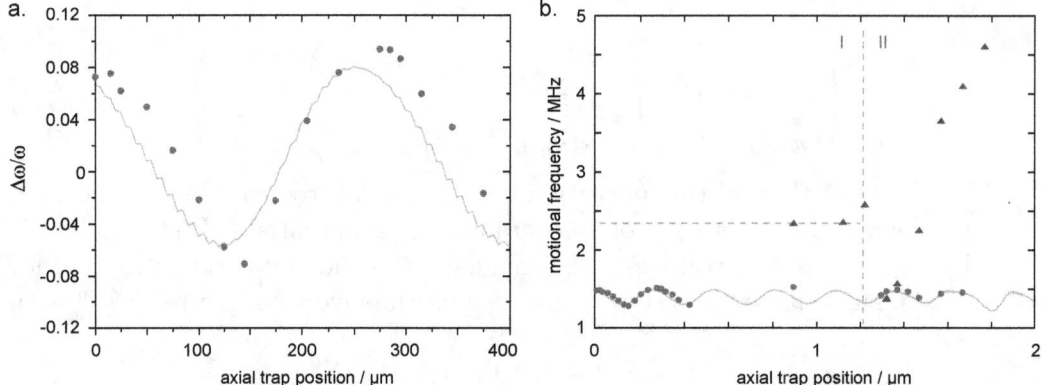

Figure 7.17: Axial trap characterization by motional frequency measurement of a single ion via quantum jump spectroscopy: (a) The numerically simulated axial motional frequencies (orange) are measured in the linear storage region for different axial positions with high accuracy (gray). The sinusoidal variation $\Delta\omega/\omega$ of the theoretical axial frequency is reproduced impressively. (b) In the storage (I) and the tapered region (II) the variation of the axial frequencies are well-reproduced. The influence of the tapered region (II) on the radial motional frequencies (triangles) depends significantly on the axial trap position compared to the storage region (I).

The measurements of the transport spectroscopy (Fig. 7.17) show spectacularly the capabilites of the extended quantum jump spectroscopy scheme for the characterization of segmented ion traps. The axial motional frequency is measured on a microscopic scale in the storage region (Fig. 7.17a) and in a more macroscopic measurement at the transition from the storage to the linear tapered region (Fig. 7.17b). To illustrate the accuracy of the delevoped algorithm for the calculation of the transport voltages, a sinusoidal variation (Fig. 7.17a) is imprinted on the theoretically predicted axial motional frequency dependent on the linear trap position (orange). The measured axial frequencies show a slight shift to the numerical simulated curve, which is explicable by an initial alignment offset. The magnification of the optical lens system forbid the simultaneous detection of the trap electrode edges and the position of the ion, so the alignment error can be estimated to $\pm 25\mu$m. The overall axial distance of the measurement starts at the middle of a control electrode pair and covers the interesting region to the adjacent pair of control electrodes. The algorithm compensates the discrete leap very efficiently and allows a smooth transport with a constant axial frequency. Severe variations on the electrode geometry are absorbed on a macroscopic scale as well (Fig. 7.17b).

The simulation and realization of a single ion transport with a dedicated axial motional frequency shows the excellent applicability of the microtrap for the investigation of scalable quantum algorithms with integrated shuttling operations. Furthermore the voltage waveforms generated for the transport spectroscopy are a perfect initial guess for the investigation of non-adiabatic single ion shuttling preserving the ions quantum state using optimal control techniques.

The measurement of a single radial frequency in the 729nm sideband spectrum for different axial trap positions show interesting features based on the change of the trap geometry (Fig. 7.17b). Starting at the end of the storage region with a constant radial frequency, a break-in after a slight increase at the beginning a the tapered region occurs. Then the radial frequency is increased strongly as expected with the transport to the narrower radial cross section of the trap. The stronger radial confinement in the tapered region is understood easily with the change of the trap geometry at constant trap drive parameters. The drop of the radial confinement at the start of the tapered region can be illustrated qualitatively with a bending of the electric field at the chamfered electrodes. The electric field is not anymore strictly perpendicular to the linear trap axis, the component parallel to the axial direction weakens the radial confinement and introduce obviously micromotion to the axial direction. This effect is persistent in the complete tapered region, but is overcompensated at some point by the stronger confinement induced by the narrowed electrode geometry.

The transport spectroscopy is realized so far with a constant trap drive voltage, but the different geometric cross sections of the trap influence the ions position in the stability region for the dynamical confinement. During the transport of a single ion from the storage to the processing region it is recommended to adjust the trap drive voltage dynamically to preserve the q-factor of the dynamical confinement in the range of 0.15 to 0.35. The single ion shuttling operation with a dynamical trap drive is tested successfully. The measurements show the influence of the tapered region on the radial confinement drastically - the persistent influence of axial micromotion generated by the taper should be considered for the application of coherent operations on single or a couple of ions. The ions should be loaded and accumulated in the storage region, then moved by a classical transport operation to the processing region, where the coherent operations on the ions quantum state should be performed.

7.5 Raman spectroscopy

The difference of the Raman scheme to the spectroscopy with the 729nm laser are a significantly enhanced Lamb-Dicke factor and the interaction of the Raman light only with the axial vibrational modes [Gul01b]. The interferometric stability of the Raman beams is important, not a small spectral linewidth compared to the requirements for the 729nm laser at all. In addition the lifetime of the Zeeman sublevels of the ground state is obviously much stronger compared to the metastable $D_{5/2}$ level.

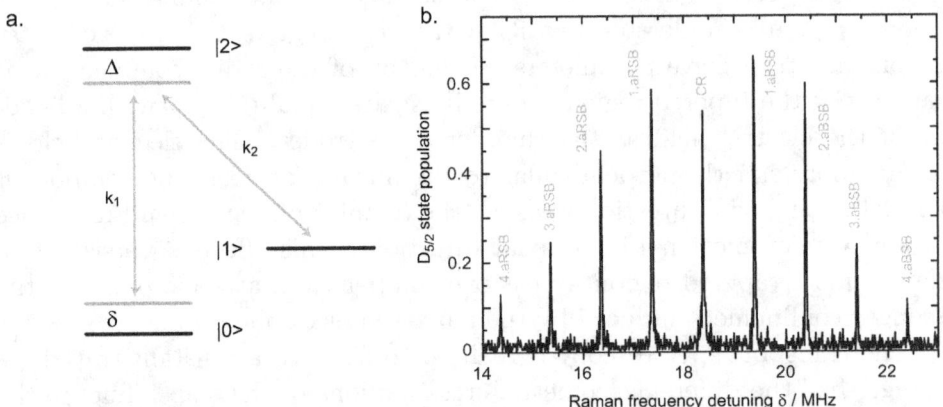

Figure 7.18: Pulsed quantum jump spectroscopy using Raman transitions: (a) The Raman levels are the Zeeman sublevels of the ground state $|0\rangle = |S_{1/2}, m = -1/2\rangle$ and $|1\rangle = |S_{1/2}, m = 1/2\rangle$. The Raman transition is detuned by Δ from the $|2\rangle = |P_{1/2}\rangle$ state. The detuning $\delta = 12.4\text{MHz}$ is determined by the frequency difference of both Raman laser beams k_1 and k_2. (b) The sideband spectrum shows axial motional sidebands only because of the effective k-vector $\tilde{k} = \tilde{k}_1 + \tilde{k}_2$ parallel to the linear trap axis. The detuning is about $\Delta = 10\text{GHz}$ and the excitation was $200\mu s$. The carrier transition (CR) is centered and axial lower (RSB) and upper motional sidebands (BSB) are measured up to the fourth order.

The Raman transition used for the sideband spectroscopy is a two-photon transition as a combination of a excitation to a Raman level and a stimulated emission as the final step. The Zeeman sublevels of the $S_{1/2}$ ground state are used as Raman levels, coherently connected by two lasers based on the third state $P_{1/2}$. The first Raman laser couples the Zeeman sublevel $|0\rangle = |S_{1/2}, m = 1/2\rangle$ to the virtual Raman level, the second Raman laser the remaining sublevel $|1\rangle = |S_{1/2}, m = -1/2\rangle$. Finally, the state $|2\rangle = |P_{1/2}\rangle$ is decoupled from the Zeeman sublevels and an effective two-level Raman interaction Hamiltonian describes the system precisely [Wu96]. The coupling is influenced by the coupling strengths Ω_1 and Ω_2 and therefore intensity dependent. The effective Rabi frequency of $\Omega = \Omega_1\Omega_2/\Delta$ is larger for a smaller detuning Δ [Hom06a]. The detuning Δ is limited by spontaneous emissions to the $P_{1/2}$ level, which acts as a loss.

The time sequence for the pulsed quantum jump spectroscopy is nearly identical to the previous sequence. Now the optical pump pulse populates the lower Zeeman sublevel m = −1/2 instead of the upper level, and the Raman transition transfers the population to the upper Zeeman sublevel m = 1/2. An external magnetic field leads to a Zeeman splitting of 12.4MHz. The detuning Δ to the $P_{1/2}$ level is about $\Delta \sim 10$GHz. The detection of a successful Raman transition in the spectroscopy section is realized with the 729nm laser combined with an robust adiabatic passage pulse.

The sideband spectrum is different compared to the quantum jump spectroscopy with the 729nm laser because of the orientation of the Raman effective k-vector. The Raman beams are tilted 45° to the linear trap axis and are perpendicular to each other, so the effective k-vector is oriented parallel to the linear trap axis and no coupling to radial modes of the trapped ion is detectable. The sideband spectrum contains the axial vibrational modes only. The Lamb-Dicke parameter is determined to $\eta = \Delta k \sqrt{\hbar / 2m\omega} = 0.356$ and is by a factor of about 5 stronger compared to the coherent coupling with the 729nm laser. The two photon process combines the advantage of a shorter wavelength, a parallel orientation to the linear trap axis and a high Rabi frequency with a simpler experimental spectroscopy setup. Analoguous to the sideband cooling based on the 729nm excitation the Raman scheme allows sideband cooling as well. The principle of realization will be similar to the discussed scheme based on the 729nm laser. An excitation on the first lower axial sideband is used to annihilate a phonon of the axial vibrational mode, cooling to the motional ground state in the end.

Chapter 8

Experiments in planar traps

The experimental results achieved with the different surface electrode ion traps represent the first trap operation and fundamental characterization of the trap geometries and the electric potentials. Starting with the technology of planar ion traps a prototype experiment was built - the trap is fabricated with printed circuit board technology (8.1). The Y-shaped trap geometry has a manifold of control segments to demonstrate algorithms for shuttling of linear ion crystals, as well as split and merge operations of linear ion chains out of two and more charged microparticles. The trap is characterized (8.1.1) and linear chains of microparticles are loaded in the trap (8.1.2). The operations for splitting and merging are investigated with two microparticles, and a transport of a single microparticle is discussed and realized with a single charged microparticle (8.1.3).

The realization of planar trap experiments with $^{40}\text{Ca}^+$ ions (8.2) is encouraged by the experience of the microparticle trapping and shuttling using surface electrode trap geometries. A simple linear planar trap with several control segments is operated and characterized using ion clouds and the trapping of a single ion (8.2.1). The motional frequencies of the ion are measured by resonant vibrational excitation for different trapping potentials and compared with numerical simulations (8.2.2).

Based on the experience with the shuttling of charged microparticles in Y-shaped planar traps and the realization of a planar trap experiment a new surace electrode Y-shaped trap is designed with a manifold of control segments for transport experiments with $^{40}\text{Ca}^+$ ions (8.3). The trap is fabricated and passed first electrical tests, and the installation of the experimental setup is completed now.

8.1 Microparticle ion trap

The microparticle ion trap was initiated as a demonstration experiment for the loading and the shuttling of single ionized microparticles respectively linear crystals of ionized microparticles. The linear planar trap is of a Y-shaped geometry, so the transport of single particles through the crossing point of the three branches is investigated. The microparticle trap acts as a prototype setup for the development of microfabricated planar ion traps for scalable quantum information science.

8.1.1 Trap operation

The Y-shaped trap and the hexagon trap are installed separately in transparent housings to shield the trapped charged microparticles from atmospheric fluctuations of the outer environment. The detection of the trapped microparticles is realized with a CCD camera via classical scattering using a helium neon laser guided parallel to the trap surface. The control software allows real-time position tracking of the microparticles to support a closed feedback loop with the programmable electrode voltages. The trap drive voltage and the multiple control voltages are supplied by a self-developed device, which provides complete remote control (5.2).

The trap drive voltage and 32 control voltages are connected with a single wire respectively a ribbon cable by feedthroughs to the segmented trap installed in the inner environment. The time-dependent waveform of the trap drive can be either sinusoidal or rectangular - the developed controller supports rectangular waveforms by switching between the positive and the inverted voltage using a transistor. Alternatively the trap drive voltage is generated by a standard function generator supporting sinusoidal and rectangular waveforms, followed by a bipolar power operational amplifier BOP-36V[1] and a self-built transformer. The number of turns on the primary coil is 50, the secondary coil of the transformer consists out of 10000 turns. However, the multiple control voltages are supplied by a 32-channel high voltage digital-to-analog converter and controlled using a serial programming interface (5.2). The unipolar output voltages for the control electrodes are limited to 295V. Under normal operating conditions the trap drive uses peak voltages from 250V up to 900V at frequencies in the range of 50Hz to 900Hz.

The organic microparticles (lycopodium) are nearly spherically with a diameter on the order of 30μm. Based on the specific gravity of $\rho = 0.74\text{g/cm}^3$ the single microparticle mass can be estimated to m = 10ng. The microparticles are deposited on a stainless steel spicule at a electric potential of 280V and loaded into the center of the trapping potential. The trap is operated during loading as a mass filter to restrict a specific charge acceptance.

---

[1]Kepco Inc., Flushing, USA

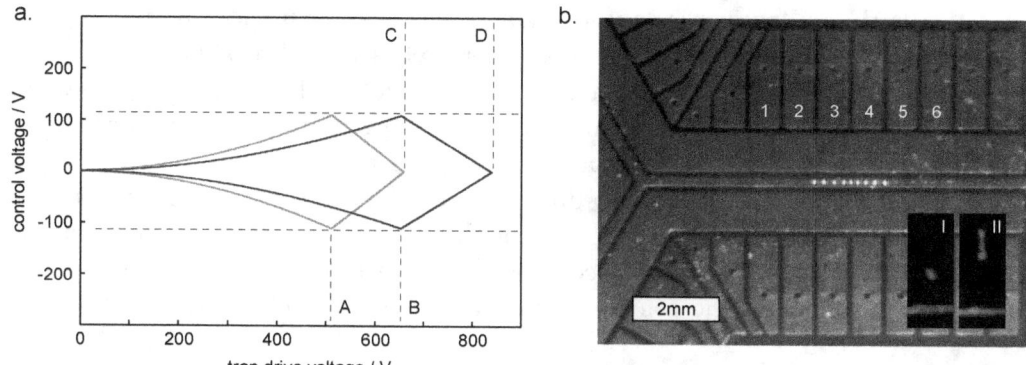

Figure 8.1: Radial and axial confinement of the microparticle trap: (a) The first region of stability is simulated numerically for a sinusoidal (blue) and a rectangular (red) trap drive. Compared to the identical shape for the control voltage (A, B), the stability diagram is narrowed (C, D) on the trap drive axis. (b) The axial confinement is shown by a linear chain of 9 microparticles. The center electrode pairs (3, 4) initiate the potential minimum, the outer electrode pairs (1, 2, 5, 6) allow shaping of the trapping potential. The inset shows a side view of a perfectly compensated (I) and an uncompensated trapped particle (II) concerning the micromotion.

The imaging of the microparticles is realized using a standard analog CCD camera combined with a framegrabber card PCI-1410[2]. The data analysis is based on particle tracking using NI-LabVIEW combined with NI-Motion. Because of the analog PAL standard the imaging update rate is limited to 25fps at an image size of 768x576 pixels. Alternatively a RTAI Linux environment[3] was used providing real-time scheduling for the voltage control. Combined with a digital IEEE1394 interfaced CCD camera DFK21BF04[4] the image readout is accelerated to an maximal update rate of 60fps at an image size of 640x480 pixels. The CCD camera uses the macro lens system Computar MLH-10X[5] with a working distance of 20cm and a magnification of 10. The real-time position tracking of the microparticles is realized with UNICAP[6]. The voltage control, image acquisition and analysis is accelerated by the real-time RTAI Linux environment. The voltage update rate is increased by a factor of 10 to nearly 10kHz and the particle tracking is speeded up by a factor of 10 to approximately 20fps.

The typical operating conditions of the segmented trap depends on the sinusoidal or rectangular shape of the trap drive voltage (Fig. 8.1), while the axial confinement is generated by the same multi-channel voltage source

---

[2]National Instruments, Texas, USA

[3]https://www.rtai.org

[4]Imaging Source Europe, Bremen, Germany

[5]CBC GmbH, Düsseldorf, Germany

[6]http://unicap-imaging.org

all the time. However, the underlying electronic circuit for the rectangular trap drive is much simpler and cost-effectively than a broad band sinusoidal amplifier up to $2000V_{pp}$ in a frequency range of 1Hz...1000Hz. Besides the technical aspect the rectangular oscillating potential leads to a narrower first region of the stability diagram (Fig. 8.1a), which is proven by a numerical simulation of the two-dimensional oscillating radial confinement of the linear trap. The Fourier series of the rectangular driving voltage $U_r(t)$ illustrates the virtually enhanced voltage amplitude by a factor of $4/\pi$, while the boundaries on the control voltage axis remain unchanged (A, B):

$$U_r(t) = U_{pp} \cdot \frac{4}{\pi} \left(\cos \omega t - \frac{1}{3} \cos 3\omega t + \frac{1}{5} \cos 5\omega t \pm \ldots \right) \qquad (8.1)$$

$$U_r(t) = U_{pp} \cdot \sum_{k=1}^{\infty} (-1)^{k-1} \frac{\cos\left((2k-1)\,\omega t\right)}{2k-1}$$

The amplitude of the sinusoidal driving voltage $U_s(t) = U_{pp} \cdot \cos \omega t$ limits the stability region of trapping (D) because of the constraint $q < 0.9$ of the q-parameter $q \propto U_{pp}/\omega^2$ used in the Mathieu equation. For the rectangular voltage waveform the stability region is narrowed by the factor of $\pi/4$ (C). The technique of trapping with a rectangular trap drive is investigated and proven even for mass spectrometry applications [Ric73, Ric75].

The micromotion of the trapped microparticles is directed perpendicular to the trap surface respectively parallel to the imaging direction (Fig. 8.1b). The orientation of the micromotion is caused by the symmetric trap geometry of the radial cross section. The time averaged detection shows a linear elongated trajectory (inset) for an uncompensated trapped microparticle. The micromotion is compensated by adjusting the voltage on the static center electrode, as long as the trajectory of the microparticle appears point shaped in the side view. Because of the particles finite dimension the micromotion is caused mainly by the finite mass. The center electrode repels the microparticels to the node of the trap drive voltage with compensation voltages in the range of 15V to 45V.

The micromotion frequency is equal to the trap drive frequency, which is proven experimentally by stroboscopic laser scattering. The motional damping of the trapped microparticles at atmospheric pressure prevents an accurate measurement of the secular and axial frequencies [Pea06]. The trap operation under vacuum pressure would eliminate the air drag. The chain out of 9 microparticles (Fig. 8.1b) is trapped by a sinusoidal trap drive with a peak voltage of $U_{pp} = 620V$ at a frequency of 280Hz. The radial pondero-motive potential is stronger compared to the axial confinement shown by the elongated linear crystal. The axial potential is generated by the control voltages $U_2 = U_5 = 180V$ and $U_2 = U_5 = 295V$ at the outer electrodes, and $U_3 = U_4 = 0V$ at the center electrodes.

8.1.2 Microparticle crystals

The loading of linear microparticle crystals with different length demonstrates the stable operation of the planar trap (Fig. 8.2). The linear crystals shown are trapped with identical parameters for the trap drive and the control voltages. The sinusoidal trap drive is characterized by a peak voltage of $U_{pp} = 660V$ at a frequency of 321Hz. The axial confinement is realized using 10 control electrode pairs in a linear branch of the Y-shaped trap. Based on the simple voltage configuration of 0V for the 4 electrode pairs at the center and 295V for the outer 3 electrode pairs at each end, the linear crystals are trapped with the middle electrode at 0V neglecting the micromotion. Assuming the critical ratio $\omega_s/\omega_{ax} \geq 0.73 \cdot N^{0.86}$ of the secular ω_s and axial frequency ω_{ax} for the linear seven-ion crystal [Ste97], the axial confinement can be estimated to at least a factor of 3.9 weaker than the radial confinement.

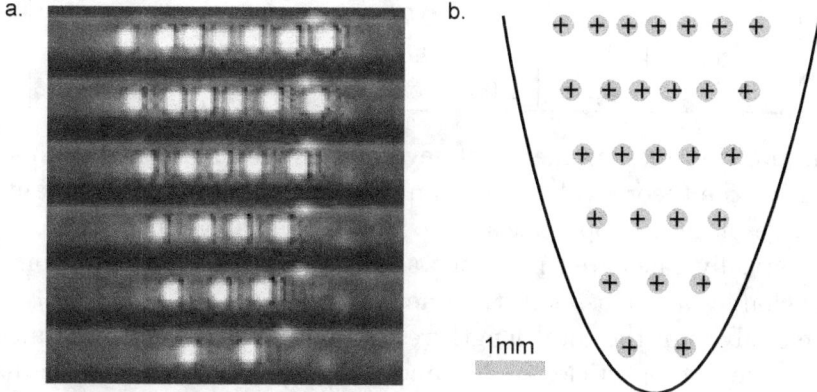

Figure 8.2: Linear crystals of charged microparticles: (a) The ion chains of 2 up to 7 microparticles trapped within an identical axial confinement are shown. (b) Compared to the theoretical equlibrium positions (gray) the distance of adjacent ions is reproduced with high accuracy, respecting slight variations of the microparticles specific mass.

The various linear crystals show the homogeneousness of the trapped microparticles (Fig.8.2a). In principle the loading process is not specific to the charge-to-mass ratio, but the trap can be operated during loading as a mass filter with a stability diagram optimized for a limited species. An averaged charge to mass ratio for the microparticles of the linear crystal is determined by the analysis of the overall length. The total length is calculated based on the equilibrium positions in the nearly harmonic axial potential. The total energy V of the linear crystal is the sum of each particle in the axial electric potential, considering the interparticle Coulomb repulsion at the distance Δz and charge Ze [Jam98]:

$$V = \sum_n \frac{Ze\alpha\, z_n^2}{2} + \sum_{n,m} \frac{(Ze)^2(1 - \delta_{nm})}{8\pi\epsilon_0 \cdot |\Delta z_{nm}|}$$

The equilibrium positions are obtained by minimizing the potential energy $dV/dz_n = 0$ numerically using a genetic algorithm. The interparticle spacing at the equilibrium is not uniform, comparably to a crystal structure of a bulk near the surface, the spacing increases at both ends of the linear crystal. The theoretical spacing between the inner particels of a 7-ion crystal is halved nearly compared to the normalized distance in a 2-ion crystal:

ions	total length	normalized equilibrium spacing					
7	3.58	0.67	0.58	0.55	0.55	0.58	0.67
6	3.19	0.70	0.61	0.59	0.61	0.70	
5	2.77	0.73	0.65	0.65	0.73		
4	2.28	0.78	0.72	0.78			
3	1.71	0.85	0.85				
2	1.00	1.00					

The position measurements of several linear crystals are in good congruity with the theoretical equilibrium positions assuming a constant mass and charge of all microparticles (Fig. 8.2b). The singular deviations from the numerically calculated positions are due to a slight distribution of the specific charges and masses in the linear crystal.

The analysis of the total length results in an averaged charge measurement for the microparticles of the linear crystal. Starting with a chain of multiple particles, the microparticles are removed successively out of the chain. The total length is measured to achieve the average charge: Following the ansatz of [Jam98], the detected particle positions z_n are compared with the dimensionless equilibrium positions u_n, which are calculated by a genetic algorithm and are consistent with the data in [Jam98]. The length scale $l = Ze/4\pi\epsilon_0\,\alpha$ is the scaling factor for $z_n = l \cdot u_n$. The length scale $l = z_n/u_n$ is fitted for different sizes of the ion chain to determine the averaged charge Ze of a single microparticle. The coefficient α is numerically calculated for a given voltage configuration by the axial harmonic electric potential $\phi = 1/2\,\alpha z^2$. The axial potential defines the single microparticles potential energy $V = Ze \cdot \phi$. The total length of linear ion chains in the scaling of u_n from 2 ions up to 9 ions are: 1.26, 2.15, 2.87, 3.49, 4.02, 4.51, 4.95, 5.36. The control voltages for the axial confinement are 295V for the three outer electrode pairs at each side and 0V for the four electrode pairs at the center. The shape of the harmonic axial potential is given by the coefficient α, which is calculated to $3.1 \cdot 10^5 \mathrm{V/m^2}$ based on the voltage configuration at the specific electrode geometry.

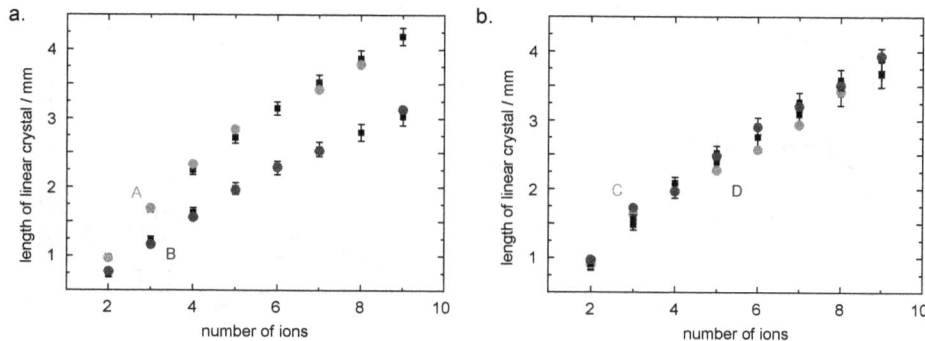

Figure 8.3: Length measurement of linear microparticle crystals: The linear chain is loaded into the trap, successively the ion number is decreased. The control voltages are constant. (a) Different trap drive parameters lead to a slightly changed axial confinement. The trap drive voltage of $740V_{pp}$ at 903Hz (A) causes a different slope compared to the linear crystal (B) at $880V_{pp}$ and 321Hz. (b) A slightly change of the averaged charge (C, D) is detectable with this technique. The trap drive is operated at $660V_{pp}$ at a frequency of 323Hz.

The parameter l is determined for four different linear ion crystals, loaded with identical parameters for ionization of the neutral microparticles (Fig. 8.3). The fitted parameter $l_A = 0.64(2)$mm results in a averaged charge of $Z_A = 5.6(5) \cdot 10^4$ec (elementary charge) and for $l_B = 0.89(4)$mm a averaged charge $Z_B = 15(2) \cdot 10^4$ec is achieved. The averaged charge of the linear ion crystals is slightly different with $l_C = 0.73(4)$mm corresponding to $Z_C = 8(1.3) \cdot 10^4$ec and $l_D = 0.69(3)$mm results in $Z_D = 7.1(9) \cdot 10^4$ec. The measurements show that an averaged charge for a linear crystal out of similar microparticles can be determined roughly based on the total length of the chain only. Furthermore the analysis shows the small variation of the measured charges for the successively loaded crystals, confirming the quality of the enhanced loading process.

8.1.3 Shuttling and splitting operations

The shuttling of single trapped microparticles is fundamental for a scalable trap operation. The continuous transport is demonstrated for a distance of 5 electrode pairs based on different time-dependent voltage waveforms (Fig. 8.4). In a simple guess the voltages are ramped stepwise showing a discontinuous shuttling, which is improved by a linear transformation on the time coordinate (Fig. 8.4b). A finer voltage interpolation combined with a higher voltage update rate results in a continuous single particle transport. The real-time position tracking of the single shuttled microparticle show the potentiality of the consistency check regarding to the numerically calculated voltage waveforms for the control electrodes.

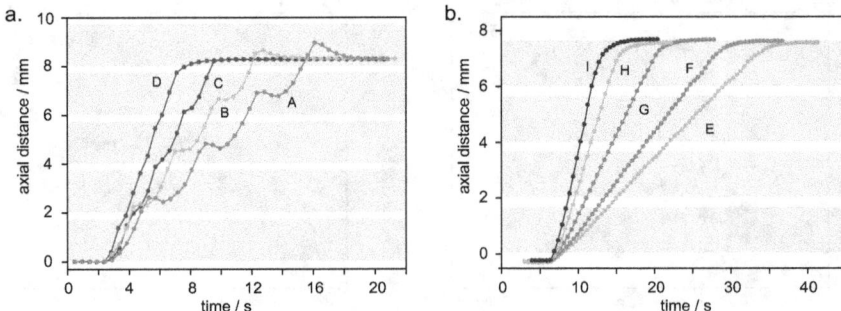

Figure 8.4: Shuttling operation with a single microparticle: (a) The six control electrode voltages of $120V, 60V, 0V, 0V, 60V, 120V$ are ramped stepwise. The trap drive voltage is $850V_{pp}$ at 50Hz with the middle compensation electrode at 3V. The delay time of 600ms (A), 450ms (B), 300ms (C) and 200ms (D) determines the transport quality. (b) The six control electrodes are ramped linearly. The initial set of voltages is $180V, 90V, 0V, 0V, 90V, 180V$ with a trap drive of $280V_{pp}$ at 70Hz. The middle electrode is set to 0V. The delay time of 800ms (E), 640ms (F), 400ms (G), 240ms (H), 160ms (I) increases the shuttling speed of the monotonic trajectory and demonstrates the optimized transport quality.

The splitting and merging operation of two ions is fundamental for the application of scalable quantum algorithms at the qubit system. The typical axial confinement of the linear ion crystal out of two ions is adjusted to place a single control electrode pair in between. This electrode pair (dark gray) allows a deterministic splitting by increasing the voltage (Fig. 8.5). The trajectories during splitting and merging operation are correlated strongly with damping effects originated by the residual pressure. However, the principle is shown clearly and further enhancement should focus on the design of the center electrode for a narrowed axial electric splitting potential.

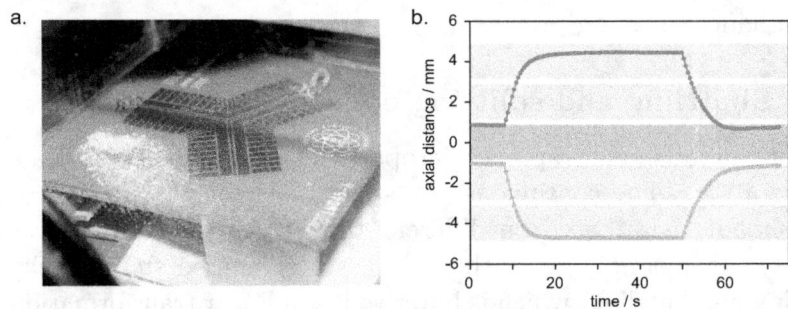

Figure 8.5: Splitting and merging of a linear two-ion crystal: (a) Two ions are confined in a harmonic axial potential. The trap drive is $900V_{pp}$ at a frequency of 50Hz. (b) The real-time particle tracking shows the influence of the increasing and decreasing double well potential on the damped motion.

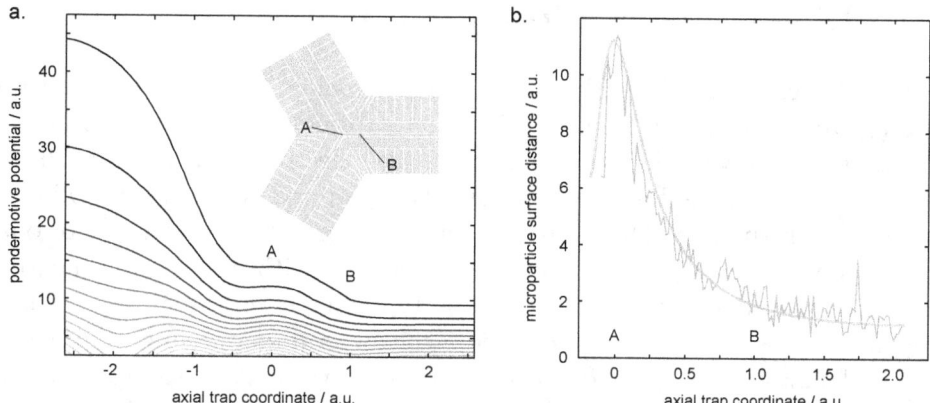

Figure 8.6: Axial ponderomotive potential at the cross: (a) The numerical simulation of the trap drive potential at the axial plane of symmetry shows the uninfluenced linear trap characteristics (on the right of B). The maximum of the trap drive bump at the axial direction is centered at the cross (A) and decays into the linear regions (B). (b) The theoretical distance of the microparticle to the trap surface for the shuttling throughout the cross (gray) is proven experimentally with high accuracy (orange). The detection of the bump confirms the existence of axial rf potentials directly and shows the accuracy of the numerical simulations.

The axial confinement in the linear regions of the Y-shaped microparticle trap is purely static (Fig. 8.6). Towards the cross (A) the third trap drive electrode located perpendicular to the linear trap axis influences the static axial trap potential strongly. The oscillating electric field component parallel to the axial direction results in a ponderomotive potential with increasing slope to the trap center (B). The transport of a single trapped microparticle to the cross starting at the linear region and finally to a different linear branch is characterized by a bump of the ponderomotive potential with its maximum at the crossing point (Fig. 8.6a).

In addition the third electrode induces axial micromotion. The strength of the axial micromotion is maximized at the increasing slope between (B) and (A) and vanishes at the cross (A). The special case at the crossing point is due to the 120°-symmetry of the geometric design and is illustrated by an enlarged plateau of the ponderomotive potential.

The potential bump of the oscillating field is detected directly by measuring the distance of the microparticle to the trap surface at a transport throughout the crossing point (Fig. 8.6b). With a constant voltage at the middle electrode the microparticle reproduces the axial equi-pseudopotential directly. The theoretical curve from numerical simulations is confirmed by the measurement with excellent accuracy. For this measurement the real-time Linux based detection was used, providing a higher data acquisition speed shown by the single experiment.

8.2   Planar microchip ion trap

The initial experiments with the planar microchip trap demonstrate the loading and storage of single ions using microfabricated single-layer electrode structures. However, the operation of surface electrode ion traps is shown successfully by [Sei06, Lab08a, Wan08]. The loading and Doppler cooling at room temperatures [Sei06] as well as the sideband cooling to the motional ground state at cryogenic temperatures [Lab08a] with outstanding heating rates are realized so far, but the availability of single ion shuttling in planar trap geometries is still an open question. The operation of a microfabricated linear planar trap is a good starting point to verify the accuracy of the numerical simulations by the measurement of axial motional frequencies. The Y-shaped planar trap as an analogous design of the microparticle trap is fabricated and characterized, but still waits for operation. The multiple control segments in the linear regions and the cross at the center will allow extensive investigations of ion shuttling inside these trap designs and prove the scalability of surface electrode traps in general.

8.2.1   Trap operation and cold ion crystals

The microfabricated planar trap is characterized by axial motional frequency measurements with ion clouds and the storage of single ions. The atomic calcium beam is aligned parallel to the trap surface, the neutrals are ionized using resonant two-color photoionization. The measurements reported here are realized at the center of the microtrap, the lasers for photoionization, Doppler cooling and repumping are guided about $\sim 245\mu$m above the trap electrodes crossing the thermal effusive beam of neutral calcium.

Typically the trap is supplied with rf voltages of U $= 300$V$_{pp}$ at a frequency of $\Omega = (2\pi)18.34$MHz. The trap is operated in the first stability region, the q-parameter is estimated using numerical simulations to q $= 0.27$. The pure radial confinement without any static voltages results in a radial trap depth of 110meV. Compared to a thermal energy of 25meV at an ambient temperature of 300K, the radial confinement by the planar electrode geometry is suitable for ion trapping after initial Doppler cooling. Based on numerical calculations, the trap center at the rf node is located 246μm away from the electrode surface. The theoretically predicted focus of the optical detection is proven experimentally with stored ions.

The operation of the oven, the photoionization, the Doppler cooling and the fluorescence detection are realized similar to the microtrap experiment based on the same characteristic parameters. The oven is operated at currents of 3A continuously, the first resonant step of photoionization to a Rydberg level is realized with the 423nm laser, the second non-resonant laser at 390nm is aligned parallel and excites the neutrals near the ionization continuum for a subsequent field ionization in the Paul trap. The Doppler cooling of the trapped ^{40}Ca$^+$ ions is achieved at a laser power of 1mW for the ion clouds and < 1mW tuned near to the resonance for Doppler cooling of the single ions. The weak focus of a diameter about 80μm respects the lateral extent of the beam near the side of the trap chip to avoid backscattering. The 866nm laser aligned parallel to the laser at 397nm is used for repumping from the D$_{3/2}$ level. The fluorescence detection of the trapped ions near 397nm is identical to the microtrap setup, but the optical lens system is oriented vertically and perpendicular to the trap surface. The fluorescence is detected with the CCD camera exclusively.

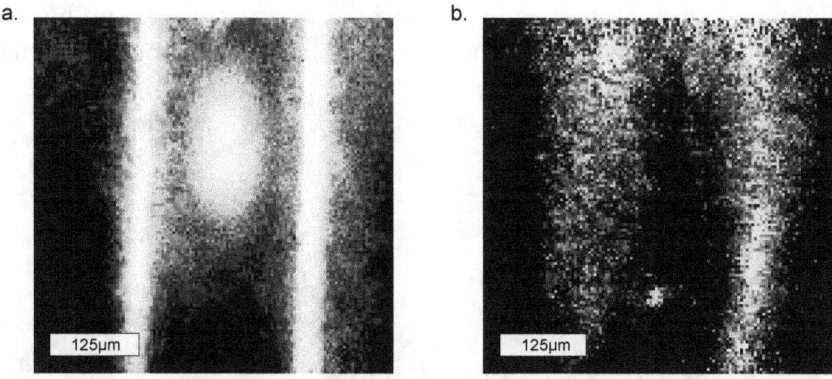

Figure 8.7: Fluorescence images of trapped ions in the planar trap: (a) The trapped ion cloud consists out of 9 ions and is centered on the middle electrode 230(20)μm above the trap surface. The scattered light at 397nm shows the ions and the gaps between the middle and the rf electrodes. (b) The single ion is stored at nearly the same distance from the trap surface. The reduced straylight is caused by a stronger focused and attenuated 397nm laser.

The fluorescence measurements of the trapped ^{40}Ca$^+$ ions (Fig. 8.7) are limited by the storage time. The trap is operated at a high-vacuum pressure of 10^{-9}mbar, so the ion loss due to scattering because of the background pressure restricts the storage time to several seconds without Doppler cooling. The ion storage time is extended using Doppler cooling by several orders of magnitude. Because of the symmetric radial cross section the micromotion is directed mainly perpendicular to the trap surface. A component of the 397nm laser parallel to the micromotion is not existent, so the vertical principal axis of motion is not affected by Doppler cooling which limits the storage time furthermore. Even the micromotion compensation by the middle control electrode is not detectable by the change of the fluorescence signal compared to the microtrap setup. The middle control electrode allows decreasing the effective distance of the trapped ion to the trap surface, so the influence of the patch charges localized at the insulators caused by the electrode gaps is different. Because of the dominating pressure-limited storage time, this effect is not detectable at this stage.

8.2.2   Trap characterization

The trap is characterized by the measurement of the axial motional frequencies for several different strengths of axial confinement. The ponderomotive potential generating the radial confinement remains constant, while the axial confinement is adjusted individually by the control voltage of the electrode pair at the center of the static potential, at which the ions are located (Fig. 8.8). The motional frequencies are determined using the technique of resonant vibrational excitation of the ions. Therefore a single electrode of a pair near the center of the axial electric potential is supplied with a small fraction of a sinusoidal voltage at a specific frequency. By scanning the frequency of the oscillating probe potential the ions get resonantly with the static potential at the axial frequency, and a dip in the fluorescence signal of the trapped ions dependent of the voltage frequency is detected.

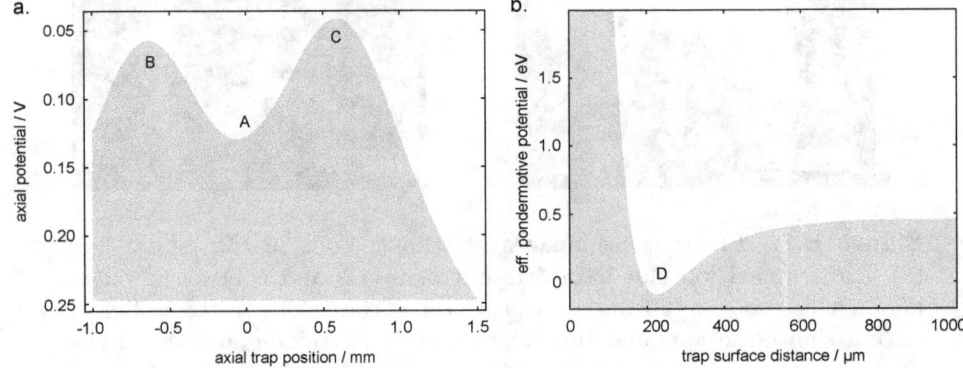

Figure 8.8: Numerical simulated planar trap potentials: (a) The pure static axial confinement at the linear trap axis is shown for a specific voltage configuration. The 10 control voltages are 0V, 10V, 10V, 0V, −5V, −2V, 6V, 10V, 10V and 0V. The non-segmented center electrode is at a voltage of −0.2V. The ion is located at the center (A), and the weak axial potential is spread about 1.2mm (B,C) with a potential depth (A,B) of about 0.6eV. (b) The trap drive voltage of $140V_{pp}$ at 18.543MHz combined with the static control voltages results in a strong effective radial confinement of about 0.5eV perpendicular to the surface. The ion is trapped at the rf node $246\mu m$ above the electrode surface (D).

For a specific voltage configuration, the asymmetric shape of the axial electric potential is caused by the irregular number of electrodes at each side of the ions position (Fig. 8.8a). The axial motional frequency is increased by lowering the voltage at the central electrode pair located at zero position. The effective ponderomotive potential (Fig. 8.8b) is obtained combining the ponderomotive potential from the trap drive and the static electric potentials of the surrounding control electrodes, especially the non-segmented center electrode. A negative voltage at the center electrode increase the effective radial confinement in the vertical direction. The pure ponderomotive potential leads to a trapping depth of about 110meV, a center electrode voltage of $-0.2V$ increase the effective radial confinement neglecting all micromotion. The voltage configuration for the trap drive and the control electrodes is optimized for the loading and the storage of the ion cloud.

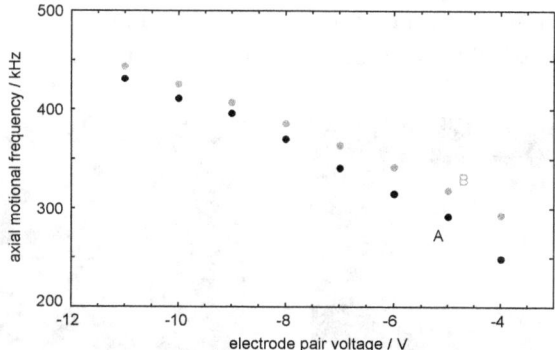

Figure 8.9: Characterization of different axial confinements of an ion cloud by the measurement (A) of their specific vibrational resonance frequency compared to numerical simulations (B). The voltage configuration 0V, 10V, 10V, ΔU, U, $-2V$, 6V, 10V, 10V, and 0V is varied by the parameter U only, the middle electrode is supplied with a static voltage of $-0.2V$. The amplitude of the sinusoidal voltage ΔU for vibrational excitation is $1meV_{pp}$.

The trapped ion cloud is used for the characterization of the electric fields by measuring the axial motional frequencies for different voltage configurations (Fig. 8.9). The axial motional frequency is varied by the voltage U applied to the central electrode pair. The measurement (A) shows a stronger axial confinement for a decreasing voltage U, which corresponds to the numerical simulations (B). The shape of the curve is reproduced with very high accuracy, the obvious offset is presumably due to imperfections of the numerical simulation. The congruity of the axial motional frequency measurements with the numerical simulations illustrate, that the mechanism of planar trapping is well-understood and the simulations of the electric fields are a convinient initial guess for ion shuttling.

8.3   Planar Y-shaped ion trap

The planar Y-shaped ion trap is derived from the design of the microparticle trap. The electrode geometry of the symmetric cross section at the linear trap region is optimized based on numerical simulations and the experience of the microparticle trap operation. A single loading zone with an asymmetric cross section is integrated to tilt the orientation of both principal motional axes to the electrode surface and optimize the Doppler cooling.

a.

b.

Figure 8.10: Planar Y-shaped ion trap: (a) All rf and dc electrodes are controlled independently via the chip carrier. The trap electrodes are connected with ball bonding to the carrier contact pads. The trap is mounted on a copper base to heighten the trap surface for optimal laser access. (b) The trap is installed on a filter board providing a cutoff frequency of 1MHz for each control electrode. Two resistively heated stainless steel ovens generates thermal effusive beams of neutral calcium.

The trap is installed in a UHV compatible ceramic chip carrier analogously to other experimental setups (Fig. 8.10). The three symmetric linear regions are connected to the trap center, 20 electrode pairs are available for ion shuttling and 5 electrode pairs are used for the axial control of the asymmetric loading region (Fig. 11.6). Each electrode of a pair is supplied with an independent static voltage. The self-developed multi-channel voltage supply from the microtrap setup is used. The rf electrode bars for the radial confinement of the trapped ions and all dc electrodes are connected with ball bonding to the chip carrier contacts.

The trap fabrication is investigated using electron microscopy (Fig. 8.11). The high aspect ratio of 2.4 regarding to the trap electrode thickness and the electrode gap will reduce the influence of patch charges on the trapping potentials. In a test setup the unexspected high breakdown voltage on the order of 650V ... 850V for a electrode gap of 4μm at a pressure in the range of 10^{-8}mbar shows the applicability for the rf trap drive to a couple of hundred volts. The surface roughness is improved regarding the linear planar microtrap.

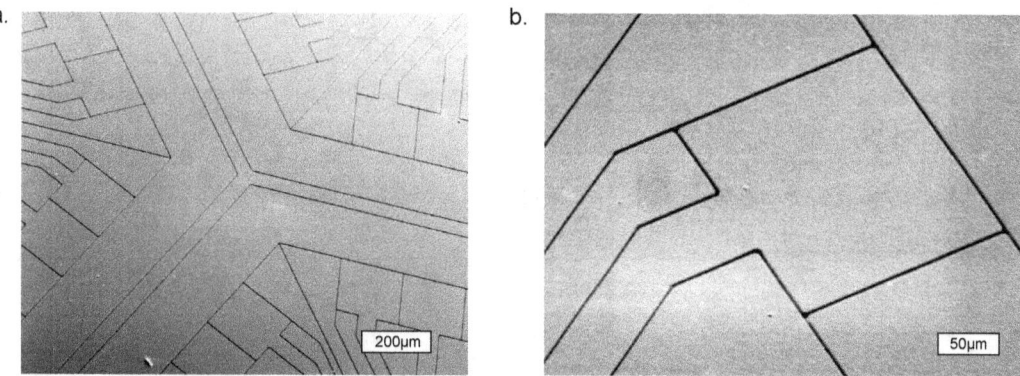

Figure 8.11: Electron microscopy of the planar Y-shaped trap: (a) The high quality of the single-layer lithography process and the clear separation of the gold electrodes are shown at the main trap region. The planar trap consists out of gold electrodes with a thickness of 4μm and a spacing of 2.3μm. (b) The surface roughness is estimated to less than 10nm, showing well-defined edges of a single control electrode near the cross.

The radial cross section at a linear branch of the planar trap shows a symmetric design with a 40μm wide non-segmented middle electrode for biasing the height of the trapped ions for micromotion compensation (Fig. 8.11a). The adjacent rf electrode bars have a width of 175μm and are surrounded by various segmented control electrodes with a constant width of 180μm. The control electrodes has a nearly squared shape (Fig. 8.11b), the length is 190μm. The control electrodes are supplied with voltage by 50μm wide striplines, that are surrounded by a large ground area. The linear symmetric branches are merged in a cross at the trap center, the asymmetric loading region is attached to a single linear branch and not shown. At the radial cross section of the asymmetric loading region the 60μm wide center electrode is surrounded by a 76μm wide rf electrode and a 375μm wide control electrode on the one side, in the other direction the rf electrode has a width of 360μm and the outer control electrode a width of 180μm. The length of the middle electrode pair in the loading region is 250μm. In axial direction this electrode pair is surrounded by two additional control electrode pairs in each direction, which transfer the asymmetric cross section to the symmetric linear geometry smoothly (cp. Fig. 11.6).

Based on numerical simulations of the electric fields the ponderomotive potential at the symmetric and asymmetric radial cross sections is shown. The trap parameters for operation are calculated with a stability parameter of q = 0.3. The geometric factor of $\alpha = 2.8 \cdot 10^7 \mathrm{m}^{-2}$ is determined at the radial cross section of the symmetric region. A trap drive voltage of U = 280V$_{\mathrm{pp}}$ at a frequency of 40MHz yields to a radial trap depth of 350meV for a pure dynamical confinement of the trapped ion. The rf node is located 72μm above the surface electrodes in the symmetric geometry, at the asymmetric loading zone the rf node is shifted $21\,\mu$m towards the smaller rf bar, the height is increased slightly to $77\,\mu$m. The principal axes of motion are oriented parallel and perpendicular at the symmetric cross section, so Doppler cooling of the vertical motion is excluded with a parallel to the trap surface aligned laser. The asymmetric electrode design effects a tilt of 5° for the motional axes to the trap surface to enables Doppler precooling for the vertical direction.

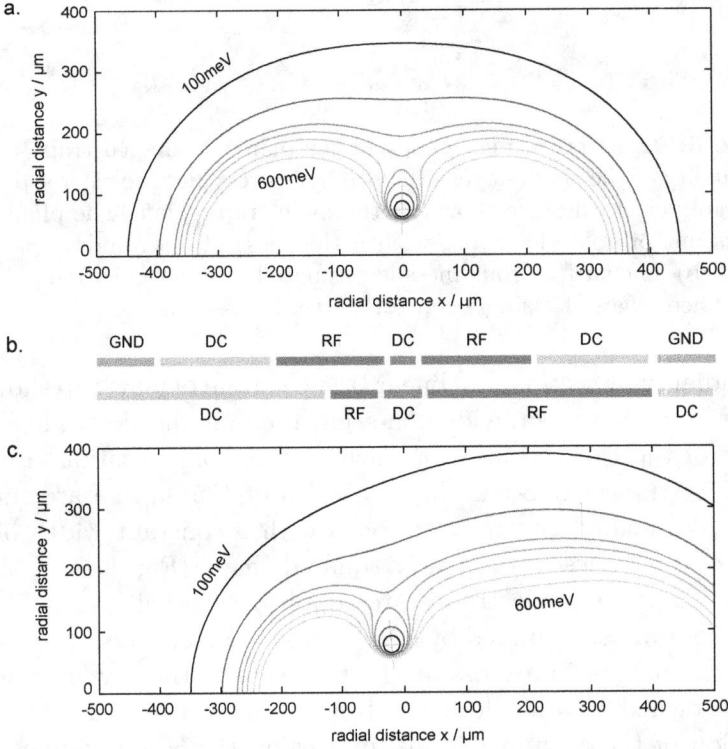

Figure 8.12: Numerical radial field simulations of the fabricated Y-shaped ion trap. The pseudopotential at the transverse cross sections of the symmetric (a) and asymmetric region (c) show the radial trap confinement. The equi-pseudopotential lines at 0.1eV, 0.2eV, 0.3eV, 0.4eV, 0.5eV, 0.6eV and 0.345eV illustrate the trapping of the ion. The trap depth of > 0.345eV for both cases (a) and (c) under normal operating conditions of 140V trap drive at 40MHz frequency is optimized by numerical calculations. The different electrode designs (b) show the origin of the tilted electric field.

The principle of axial confinement is analogous to the microparticle and linear planar trap. Several static potentials generated by single pairs of control electrodes are added to a static confinement with motional frequencies of a couple of MHz. The overlap of adjacent axial potentials is comparable to the microtrap, so the numerical calculations for a smooth transport can be transferred directly to the Y-shaped planar trap. The trap is fabricated yet, breakdown tested and just installed for the first trapping of ^{40}Ca$^+$.

Chapter 9

Summary and Conclusion

The numerical simulation, development, assembly, operation and characterization of new scalable microfabricated ion traps for quantum information science was the aim of the work.

The numerical simulations of the segmented linear ion traps contain the optimization of the electrode design calculating the electric fields using finite element methods and boundary element methods. Deviations in terms of higher multipole terms from an ideal time-dependent quadrupole potential can be eliminated efficiently by an adjustment of the trap design. The anharmonic hexapole coefficient vanishes for a special set of design parameters in radial and axial direction.

The theoretical investigation of the motional heating during single ion shuttling is of special interest for the scalable operation of the microtrap. The transport of a single ion from a specific electrode to the adjacent pair increase the external energy by heating effects. From a simple guess the time-dependent voltages for shuttling are optimized using optimal control techniques to minimize heating. The heating is classified like anharmonic dispersion and parametric heating. The optimization shows that parametric heating can be eliminated systematically and the anharmonic dispersion is reduced by orders of magnitude in classical phase space displacement. The simulation is done on a non-adiabatic timescale, which separates the shuttling limits from the trap characteristics. Transport operations in the diabatic regime will play an important role for the realizing of quantum algorithms with moving qubits.

The trap development was focused on the microfabrication of linear ion traps with a multi-layer and a planar design. A modular three-layer microtrap with 32 electrode pairs in two geometrical different adjacent zones was constructed. One trap region is dedicated for the loading of ions and the storage of qubits, the other trap region serves for the processing of quantum information. The trap is assembled in a commercial ceramic chip carrier for standardized electrical connectivity. A central constraint in the design was the limitation of the control voltages to standard output voltages from commercial digital-to-analog converters in the range of ± 15V. For the trap operation a multi-channel voltage source was developed, allowing transport close to the beginning of the diabatic regime. The modular three-layer design was realized in the foresight of hybridization to implement fibre optic in the intermediate trap layer. This trap design is the reference design in the European research program MICROTRAP, and is delivered to several European project partners for the application of quantum information science with microtraps.

The design of multi-segmented planar traps was started with traps for microparticles fabricated with printed circuit board technology. A Y-shaped planar trap was developed in a three-layer design for separating the trap electrodes with a intermediate ground layer from the electrical connectors. A programmable voltage source for trap control was designed, supplying 32 voltages up to 350V and a trap drive with peak voltages of 1000V at frequencies up to 1kHz. Based upon the microparticle planar trap a Y-shaped planar ion trap with a manifold of 70 control electrodes was designed, fabricated, tested and installed in an UHV environment.

A complete new setup was built for the microtrap experiments. Laser sources and optics were installed and the vacuum environment designed and installed with the trap. The diode lasers at 397nm, 866nm and 854nm are locked to linewidth of 1kHz using self-fabricated cavities, the quadrupole transition laser at 729nm was stabilized with a high finesse cavity to a linewidth on the order of 100Hz.

The operation of the three-layer microtrap is demonstrated with the trapping of single ^{40}Ca$^+$ ions and linear ion crystals. The quantum state readout of the ion is realized by electron shelving with a fidelity of 99.5%. The Doppler cooled ion is placed in the Lamb-Dicke regime and located within approximately 105nm at the trap node, which was evaluated by the strength of the micromotion sideband using quantum jump spectroscopy. The quantum state preparation is done by simple single pulses or robust adiabatic passage on the quadrupole transition. The coherent quantum state manipulation between $|0\rangle = |S_{1/2}, m = 1/2\rangle$ and $|1\rangle = |D_{5/2}, m = 5/2\rangle$ is demonstrated with Rabi oscillations on the quadrupole transition and Ramsey interference experiments, which shows the enhanced contrast by one order of magnitude using power line triggering. A single ion is cooled to the ground state of motion by sideband cooling on the quadrupole transition, conceding the measurement of a heating rate of two phonons per ms in the storage zone as an upper limit. Another option for defining qubit states in ^{40}Ca$^+$ is evaluated by Raman transition on the ground state $|S_{1/2}\rangle$, using the Zeeman sublevels $|0\rangle = |S_{1/2}, m = 1/2\rangle$ and $|1\rangle = |S_{1/2}, m = -1/2\rangle$. The experimental scheme to establish an optical qubit using the quadrupole transition and a spin qubit based on the Raman transition was proven.

A new type of spectroscopy experiments for trap characterization is initiated called 'transport spectroscopy' using quantum jump spectroscopy on the quadrupole transition. Solely the laser at 729nm is focused to the point of interest, where the information of the trap should be stored from the motional sidebands. The Doppler cooling and readout scheme stays at the same trap region, the trap can be fully characterized by the successive displacement of the 729nm laser on the shuttling forwards and backwards of the ion. The axial and radial frequencies of motion are measured dependent on the axial position of the ion very precisely.

Based upon the experience with the microparticle planar trap, in which the shuttling of single ions and linear ion crystals were demonstrated, a new trap setup was built for a microfabricated planar ion trap. Because of the challenging production process the characterization of the multi-segmented Y-shaped trap with trapped ions is pending, a simpler version of a linear planar trap fabricated by N. Daniidilis and H. Häffner from the University of Innsbruck was installed in a cooperation. The trap operation was demonstrated successfully and the axial motional frequencies are measured by resonant motion excitation for different trap parameters. The frequencies are in very good congruity with the numerical simulations. The experience from the Y-shaped microparticle trap with respect to the ion shuttling beyond the cross structure is essential for operating the Y-shaped planar ion trap in future.

In the end the work presented the operation of the first European scalable microtrap which is suitable for quantum logic experiments. The manifold of trap segments is the highest number of electrodes integrated in traps for quantum information science so far. A complete new experiment, the electronics and sideband cooling on the quadrupole transition, as well using a Raman transition, were demonstrated. The development on planar ion trap technology is started with the first operation of a European surface-electrode trap in cooperation with H. Häffner from the University of Innsbruck. New planar designs are prepared and will be tested soon.

Chapter 10

Outlook

The development of scalable microfabricated ion traps with multiple segments for the realization of a quantum processor is a challenging task in quantum information science. The last decade of ion trapping in this field of research is characterized by the improvement of linear ion traps together with the optimization of the laser sources to show a variety of entanglement, quantum gate and teleportation experiments. Belonging to the technical direction the trap dimensions were decreased due to the application of microfabrication techniques, a manifold of trap electrodes were integrated for the precise control of the ions motion and the ion microtraps resembled the appearance of integrated electronic devices. Various fabrication techniques for multi-segmented microtraps were tested and the required electric fields are well-known by precise numerical calculations. It was a time period of trap optimization and the experience of 'how to build an ion trap' is exhausted at this stage.

The trap technology is well developed now and the efficiency of different quantum algorithms is proven and optimized - but each component goes ahead in their own nutshell up to now. The active integration of characteristic features of state-of-the-art microtraps - the availability of independent trap zones and the shuttling operations in between - in the sequence of a quantum algorithm is still missing. It is a question of coherence, will the coherence be preserved after the ion transport, or not? Finally, the experiment will answer this fundamental question for a scalable quantum computer based on shuttling operations.

Theoretical investigations of the transport optimization are shown in this work - numerical simulations point out that ion shuttling should be possible beyond the adiabatic regime without phonon heating [Sch06]. These experiments are scheduled in the short term and will lead to feedback algorithms for optimal transport intermediately. In the long term the shuttling of ions should be integrated in quantum algorithms, i.e. the successive entanglement of ion pairs consisting out of adjacent ions from a large ion chain.

Not only the quantum information side is a challenging task, even further development of segmented microtraps to highly integrated optoelectronic devices is of utmost significance for scalable quantum computing: The combination of electric fields with light sources provides the electrodynamics for trapping and shuttling as well as the light for adressing and readout bounded to independent trap regions. The integration of optical components to the fabrication process of the microtraps will lead to a simple optical environment and enhance separated functional trap regions. In the end the hybridization to optoelectronic devices will accomplish the integration of diode lasers including the microoptics in the trap assembly.

The functional design of the three-layer trap as the first European microtrap developed and tested in this research work is an outstanding candidate for future optoelectronic integration within the next years. The spacing of the three-layer design encourages the integration of fibres for guiding the lasers (Fig. 10.1), because the fibers are aligned automatically to the ion and the fibres are covered by the electrode layers. The electric field for trapping remains symmetric and the influence of patch charges by the insulating fibres are minimized. Some special trap regions could be established, i.e. some Doppler cooling regions and some electrode segments which provides qubit manipulation. The concept integrates the external laser sources in the trap design, and the ions are shuttled to the functional zones.

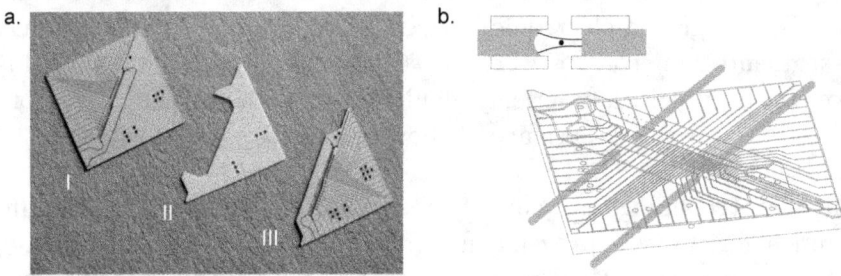

Figure 10.1: Fibre integration in a future optoelectronic microchip trap: (a) In the base system of bottom layer (I), spacer (II) and top layer (III) the spacer is modified for optics integration. (b) The fibres or fibre cavities (gray) are enclosed by the electrode layers (red, green), the inset shows the advantage of two electrode layers compared to three layer of electrodes.

In the short term fibers with microlenses on top could be mounted between the electrode layers of the trap (Fig. 10.1a), replacing the ceramic spacer. The fabrication of the microlenses is realized by shaping the end of the fiber, which is a well-known technique [Pre92, Thu03]. A tight focus in front of the fiber will minimize the scattering light of the Doppler cooling laser and therefore the background signal of the detection. The standard fibre core of 125μm is equal to the thickness of the spacer, so the trap geometry in particular the shape of the electric field is conserved.

In a more advanced scheme of the microtrap as an optoelectronic device two fibers are integrated oppositely as microscopic optical high-finesse cavities (Fig. 10.1b). Neutral atom experiments using mirror cavities [Mau05, Boc04] and recently fibre cavities on an atom chip [Col07] show the capabilites of quantum electrodynamics measurements, while the integration of fibre based optical cavities in ion microtraps for quantum information science is an open task. The fibre ends will be approached to several tens of μm to reach the strong coupling limit. The readout of the qubit information can be realized in a cavity quantum electrodynamics experiment by shuttling single qubits throughout the cavity. The quantum state of the ion will be determined by this non-demolition quantum measurement using the cavity probe field. The interaction is oriented strongly to the cavity field, which will decrease the readout time in contrast to the electron shelving technique with arbitrary spontaneous photon scattering. In this standard readout scheme using a dipole transition only a fraction of the spontaneous scattered photons like 1/1000 is collected. This limits the readout time of the qubit and the speed of a future quantum computer perspectively, which can be enhanced by the cavity readout scheme.

The realization of the qubit state detection of single ions or linear ion chains by shuttling using the control voltages through the cavity mode is done by the measurement of the cavity transmission. The fibre cavity is tuned to the $P_{1/2} \leftrightarrow D_{3/2}$ transition near 866nm. During a single ion transit the qubit state $|0\rangle = |S_{1/2}\rangle$ is coupled via Doppler cooling near 397nm to the $P_{1/2}$ level and interacts with the cavity mode. The qubit state $|1\rangle = |D_{5/2}\rangle$ is decoupled from the levels $S_{1/2}$, $D_{3/2}$ and shows no interaction with the cavity field.

The collective cooperativity $C = g^2/2\kappa\gamma$ is estimated to $C > 1$ assuming realistic parameters: The waist radius $w = \lambda \cdot \pi/2 \cdot \sqrt{2dr - d^2}$ follows from the cavity length d, the radius of curvature r and depends on the wavelength λ of the cavity field. In a tapered region of the segmented trap the fibers can be covered by the electrode layers using a fibre gap of d $= 150\mu$m with a radius of r $= 125\mu$m. A finesse of F $= 10000$ should be achievable, the resulting waist radius of 13μm at $\lambda = 866$nm permits a cooperativity of $C = 3\lambda^2 F/\pi^3 w^2 = 4.3$. The field decay rate $\kappa = \pi c/2Fd$ results in $2\pi \cdot 50$MHz. With a spontaneous emission rate γ of $2\pi \cdot 3$MHz at 866nm the collective coupling strength $g = \sqrt{3\lambda^2 c\gamma/\pi^2 w^2 d}$ follows to $2\pi \cdot 35$MHz. The fabrication of the concave-shaped fibre ends is a standard technique using a CO_2 laser. A high reflection coating is evaporated to the fiber tips afterwards. The exemplary calculation shows that the fibre integration in a microfabricated ion trap benefits the coupling strength g, which is several orders of magnitude lower in macroscopic cavity setups due to cavity lengths of several mm. The optoelectronic integration affirms the equality of light and electric fields in such a device, the fibre cavity setup is enclosed by the electrode layers of the ion trap assembled in a commercial chip carrier.

The fabrication of such a device should be realized in the short term. The advantage of the modular microtrap design allows a variety of options for integrating other devices in the intermediate layer. Nevertheless the integration of fibre cavities in the strong coupling regime seems to be realizable also with multi-segmented planar traps: Placing a plane fiber tip perpendicular to the trap surface over an integrated curved mirror [Tru07] or - a concave-shaped fiber in front of the plane substrate - experiments in the field of cavity quantum electrodynamics are also scalable with these trap designs. Because of the beginning status with planar ion traps some progress has to occur in the lithography techniques for planar trap fabrication. The fibre cavity integration at the three-layer microtraps is an active research project and funded by the European Union in the program 'Scalable Quantum Computing with Light and Atoms' (SCALA).

Beyond future research on the implementation of ion shuttling in quantum algorithms and optoelectronic hybridization, modifications of the outer environment influence the decoherence mechanisms of the ion trap directly. Lowering the temperature increases the coherence time of the qubit significantly. Experiments have shown that the heating rate of a planar microtrap is decreased by seven orders of magnitude synchronously to cooling close to liquid helium temperature [Lab08a]. The approach to a scalable quantum processor is limited by decoherence so far - the smaller the dimensions of the trap structures, the higher the heating rates $\dot{n}$. The heating rate $\dot{n}$ of the ions secular motion is increased empirically by $\dot{n} \propto 1/d^4$, when d is defined as the distance of a trapped ion to the nearest electrode surface. Typical dimensions of the microtraps are in the range of $d = 100\mu m \dots 400\mu m$ with corresponding normalized heating rates of several quanta per ms [Lab08a]. They touches almost the $100\mu s$ gate operation limit, which is adequate for implementing most of the known quantum algorithms without phonon heating. Within the next years a reduction of the trap dimensions by a factor of 10 following a 100 times higher heating rate can be exspected, which emphasizes the technical demands of cooling for a scalable miniaturized quantum processor. Microtrap experiments in a liquid helium cryostat demonstrate the temperature dependence of the anomalous heating [Ant08], so the scalable three-layer microtrap will be operated within the next year in a cryogenic environment.

Decoherence studies are of utmost significance in the present status of ion trap quantum computing, but the quantum interaction between ions and superconductors, the coupling of ions to nanoscopic objects like single carbon nanotubes will introduce a new direction of research next to quantum information processing. More insight in the dominant heating processes is undoubted crucial [Lab08b]. A crygonic environment provide temperature dependent heating rate measurements and the coupling of i.e. Bell states to microscopic devices or surfaces. The fundamental explanation of heating in the quantum regime is still an open question.

In future, the successfully tested scalable three-layer microtrap and the scalable planar Paul trap will be employed in a cryogenic environment for

- experiments towards scalable quantum computing,

- the non-adiabatic transport of ions,

- cavity QED experiments with integrated optical fibers,

- the sensing of trap surface characteristics by the ions,

- and the coupling of an ion with a nearby sub-μm mechanical object.

With the traps designed, operated and characterized in this thesis, complex tools for advances in quantum information technology were developed towards the realization of a future optoelectronic device for quantum computing.

Chapter 11

Appendix

The additional information is focused mainly on technical developments regarding to the presented results. The two-layer microtrap is the reference design in an European research collaboration towards the realization of a quantum computer (11.1). The accurate operation of the different traps, the multi-layer microtrap, the microparticle ion trap, the planar microchip and the planar Y-shaped ion trap is provided by the pin assignments of the electrical vacuum feedthroughs referring to the trap electrodes (11.2). The fabrication of the two-layer microtrap requires a faultless coating process, which was developed on the basis of a vacuum layer deposition with a rotating sample (11.3).

The operation of microtraps can be standardized using industrial components. The microfabricated ion traps are mounted newly using ceramic chip carriers, new developments will focus on the fast exchange of the microtraps. The design of a ceramic socket for the chip carriers allows the fast exchange of the traps and provides an arbitrary number of electrical contacts, which is a significant improvement in terms of the operation of scalable microtraps for quantum information science (11.4).

11.1   Microtrap design for STREP network

The microtrap developed and tested in this research work was delivered as the reference design to the research groups of the European MICROTRAP STREP network[1]. The trap was fabricated and delivered as a kit for self-assembly the the consortium members (Fig.11.1). The multi-layer trap (f) consists out of two microstructures (a,b and c,d) separated by a blank ceramic spacer (e). The backside of the top layer (b) and the front of the bottom layer (c) are directly visible for the ion.

Figure 11.1: Microtrap design distributed in the STREP network

---

[1]http://www.microtrap.eu

11.2 Trap designs and pin assignments

The pin assignments refer each pin of the four DB25-type electrical vacuum feedthroughs at the trap flange to a dedicated trap electrode or ground connection. The pin numbering of the plugs follows the standard numbering with view direction from the air side on the mounting flange of the multi-layer ion trap or planar ion trap.

11.2.1 Multi-layer microchip trap

The electrode allocation scheme shows the individual connection of 70 dc electrodes (Fig. 11.2). The top wafer electrodes (outline) are numbered with digits less than 100, the bottom wafer electrodes (gray) are named by digits more than 100. The storage region is located at DC1...DC9 and DC101...DC109 (9 segment pairs). The transfer region is numbered as DC10...DC12 and DC110...DC112 (3 segment pairs). The processing region is located at DC13...DC32 and DC113...DC132 (19 segment pairs).

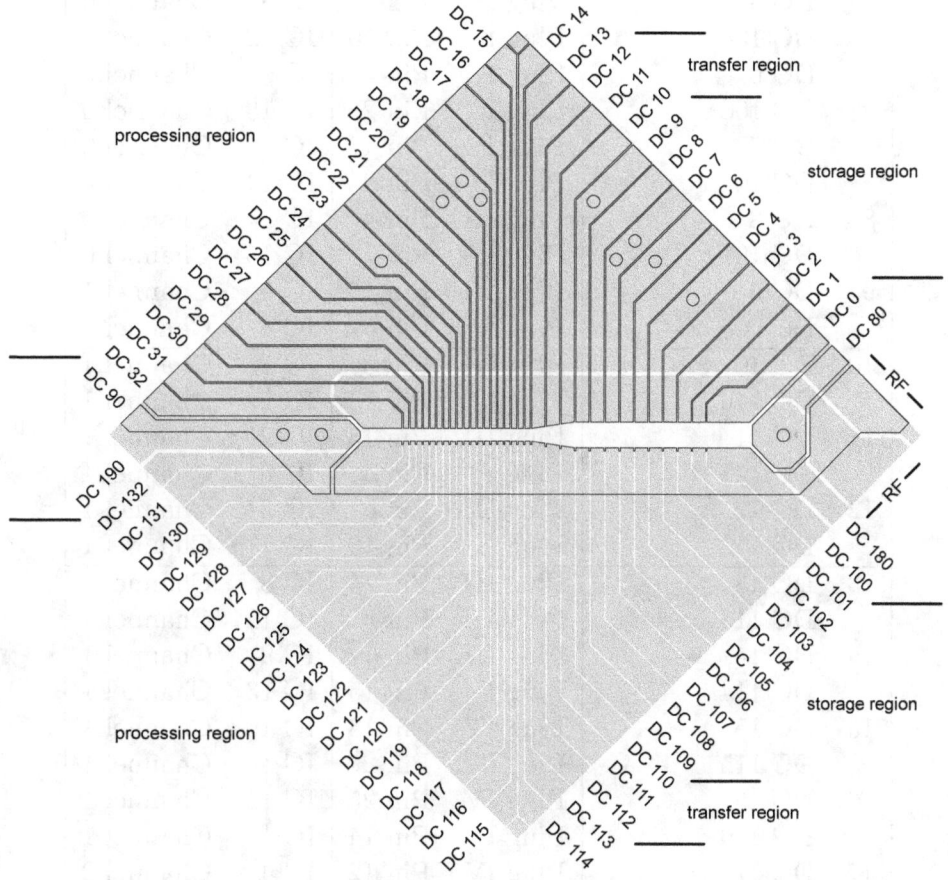

Figure 11.2: Electrode assignment of the microchip ion trap

Different electrode regions are defined using a sophisticated scheme for the assignment of the individual channels of the multi-channel digital-to-analog converters. Each electrode of a region is supported by a different digital-to-analog converter to obtain the full sampling rate of the analog output. The serial data protocol would decrease the total sampling rate if several channels of the same converter has to be controlled simultaneously.

Electrode Assignment		Pin Assignment		DAC channel	
1	DC 1	Plug III	Pin 12	IC 0	Channel 0
	DC 101	Plug I	Pin 12	IC 4	Channel 0
2	DC 2	Plug III	Pin 24	IC 1	Channel 0
	DC 102	Plug I	Pin 24	IC 5	Channel 0
3	DC 3	Plug III	Pin 11	IC 2	Channel 0
	DC 103	Plug I	Pin 11	IC 6	Channel 0
4	DC 4	Plug III	Pin 23	IC 3	Channel 0
	DC 104	Plug I	Pin 23	IC 7	Channel 0
5	DC 5	Plug III	Pin 10	IC 8	Channel 0
	DC 105	Plug I	Pin 10	IC 12	Channel 0
6	DC 6	Plug III	Pin 22	IC 9	Channel 0
	DC 106	Plug I	Pin 22	IC 13	Channel 0
7	DC 7	Plug III	Pin 17	IC 10	Channel 0
	DC 107	Plug I	Pin 9	IC 14	Channel 0
8	DC 8	Plug III	Pin 4	IC 11	Channel 0
	DC 108	Plug I	Pin 18	IC 15	Channel 0
9	DC 9	Plug III	Pin 16	IC 0	Channel 1
	DC 109	Plug I	Pin 5	IC 4	Channel 1
10	DC 10	Plug III	Pin 3	IC 1	Channel 1
	DC 110	Plug I	Pin 17	IC 5	Channel 1
11	DC 11	Plug III	Pin 15	IC 2	Channel 1
	DC 111	Plug I	Pin 4	IC 6	Channel 1
12	DC 12	Plug III	Pin 2	IC 3	Channel 1
	DC 112	Plug I	Pin 16	IC 7	Channel 1
13	DC 13	Plug III	Pin 14	IC 8	Channel 1
	DC 113	Plug I	Pin 3	IC 12	Channel 1
14	DC 14	Plug III	Pin 1	IC 9	Channel 1
	DC 114	Plug I	Pin 15	IC 13	Channel 1
15	DC 15	Plug IV	Pin 13	IC 10	Channel 1
	DC 115	Plug I	Pin 2	IC 14	Channel 1
16	DC 16	Plug IV	Pin 25	IC 11	Channel 1
	DC 116	Plug I	Pin 14	IC 15	Channel 1
17	DC 17	Plug IV	Pin 12	IC 4	Channel 2
	DC 117	Plug I	Pin 1	IC 0	Channel 2

Electrode Assignment		Pin Assignment		DAC channel	
18	DC 18	Plug IV	Pin 24	IC 5	Channel 2
	DC 118	Plug II	Pin 13	IC 1	Channel 2
19	DC 19	Plug IV	Pin 11	IC 6	Channel 2
	DC 119	Plug II	Pin 25	IC 2	Channel 2
20	DC 20	Plug IV	Pin 23	IC 7	Channel 2
	DC 120	Plug II	Pin 12	IC 3	Channel 2
21	DC 21	Plug IV	Pin 10	IC 12	Channel 2
	DC 121	Plug II	Pin 24	IC 8	Channel 2
22	DC 22	Plug IV	Pin 22	IC 13	Channel 2
	DC 122	Plug II	Pin 11	IC 9	Channel 2
23	DC 23	Plug IV	Pin 9	IC 14	Channel 2
	DC 123	Plug II	Pin 23	IC 10	Channel 2
24	DC 24	Plug IV	Pin 21	IC 15	Channel 2
	DC 124	Plug II	Pin 10	IC 11	Channel 2
25	DC 25	Plug IV	Pin 5	IC 4	Channel 3
	DC 125	Plug II	Pin 22	IC 0	Channel 3
26	DC 26	Plug IV	Pin 17	IC 5	Channel 3
	DC 126	Plug II	Pin 17	IC 1	Channel 3
27	DC 27	Plug IV	Pin 4	IC 6	Channel 3
	DC 127	Plug II	Pin 4	IC 2	Channel 3
28	DC 28	Plug IV	Pin 16	IC 7	Channel 3
	DC 128	Plug II	Pin 16	IC 3	Channel 3
29	DC 29	Plug IV	Pin 3	IC 12	Channel 3
	DC 129	Plug II	Pin 3	IC 8	Channel 3
30	DC 30	Plug IV	Pin 15	IC 13	Channel 3
	DC 130	Plug II	Pin 15	IC 9	Channel 3
31	DC 31	Plug IV	Pin 2	IC 14	Channel 3
	DC 131	Plug II	Pin 2	IC 10	Channel 3
32	DC 32	Plug IV	Pin 14	IC 15	Channel 3
	DC 132	Plug II	Pin 14	IC 11	Channel 3
	DC 0	Plug III	Pin 25	Ground	
	DC 100	Plug I	Pin 25	Ground	
	DC 80	Plug III	Pin 13	Ground	
	DC 180	Plug I	Pin 13	Ground	
	DC 90	Plug IV	Pin 1	Ground	
	DC 190	Plug II	Pin 1	Ground	
	Ground	Plug I	Pin 7, 20	Ground	
	Ground	Plug II	Pin 7, 20	Ground	
	Ground	Plug III	Pin 7, 19	Ground	
	Ground	Plug IV	Pin 7, 19	Ground	

11.2.2   Planar microparticle trap

The electrode allocation for the microparticle trap is simplified because of the high symmetric trap design and the integrated electronic circuit with a 32 channel digital-to-analog converter. The trap dc electrodes (Fig. 11.3) are controlled pairwise, each of the three identical trap regions, DC1...DC10, DC11...DC20 and DC21...DC30, consists out of 10 electrode pairs. The middle electrode DC30 is assigned separately. The extendend hexagon design (Fig. 11.4) is supplied with the same voltages.

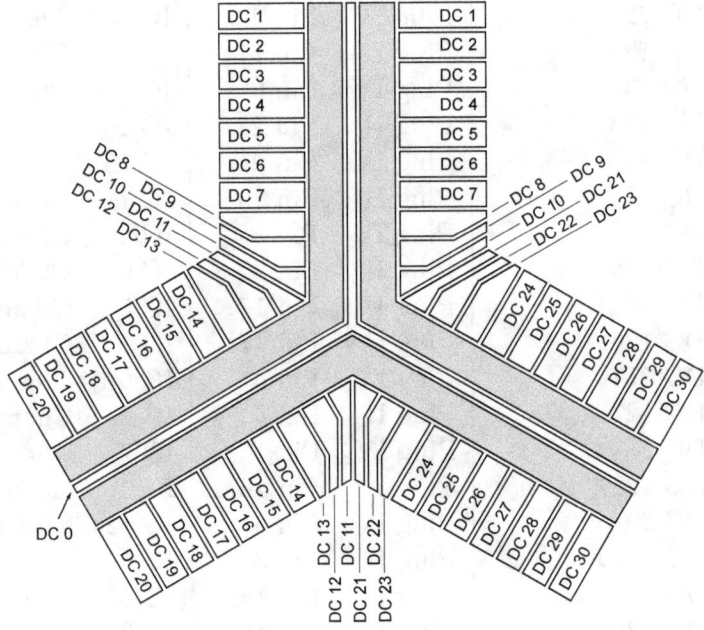

Figure 11.3: Electrode assignment of the microparticle trap

Electrode Assignment		Pin Assignment	DAC channel
1	DC 1	Pin 3	Channel 5
2	DC 2	Pin 5	Channel 6
3	DC 3	Pin 7	Channel 7
4	DC 4	Pin 9	Channel 8
5	DC 5	Pin 11	Channel 9
6	DC 6	Pin 13	Channel 10
7	DC 7	Pin 15	Channel 17
8	DC 8	Pin 17	Channel 16
9	DC 9	Pin 19	Channel 15
10	DC 10	Pin 21	Channel 14
11	DC 11	Pin 2	Channel 11
12	DC 12	Pin 4	Channel 12
13	DC 13	Pin 6	Channel 13
14	DC 14	Pin 8	Channel 4

Electrode Assignment		Pin Assignment	DAC channel
15	DC 15	Pin 10	Channel 3
16	DC 16	Pin 12	Channel 2
17	DC 17	Pin 14	Channel 1
18	DC 18	Pin 16	Channel 31
19	DC 19	Pin 18	Channel 30
20	DC 20	Pin 20	Channel 29
21	DC 21	Pin 34	Channel 28
22	DC 22	Pin 33	Channel 27
23	DC 23	Pin 32	Channel 26
24	DC 24	Pin 31	Channel 25
25	DC 25	Pin 30	Channel 24
26	DC 26	Pin 29	Channel 23
27	DC 27	Pin 28	Channel 22
28	DC 28	Pin 27	Channel 21
29	DC 29	Pin 26	Channel 20
30	DC 30	Pin 25	Channel 19
	DC 0	Pin 24	Channel 18
	Ground	Pin 1, 35	Ground

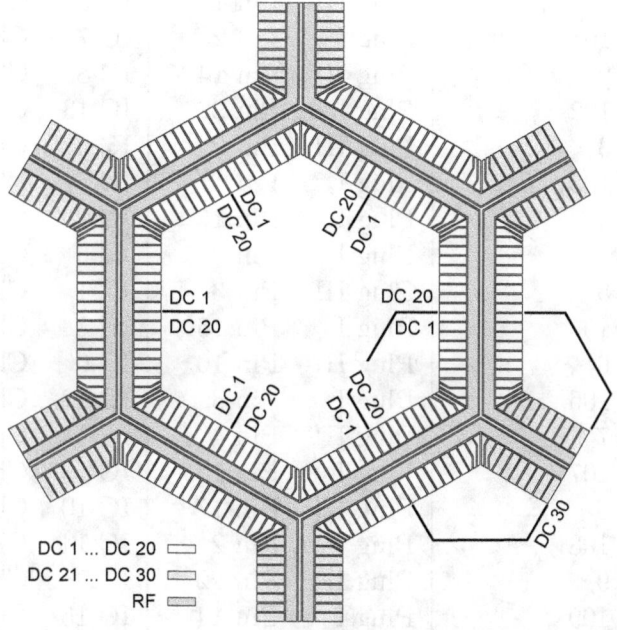

Figure 11.4: Multiplexed scalable hexagon: 180 adressable dc electrodes are controlled by multiplexing the electrode assignment of the unit cell. 20 individual dc voltage channels are adequate for shuttling. 10 dc voltages are assigned to 6 linear storage regions (gray).

11.2.3 Planar linear trap

The electrode allocation scheme shows 10 electrode pairs with a middle electrode for microcompensation (Fig. 11.5). The 21 dc electrodes are voltage controlled separately, the outer segment pairs are named DC1...DC110, the center dc electrode DC0 is enclosed by the rf electrode bars (gray).

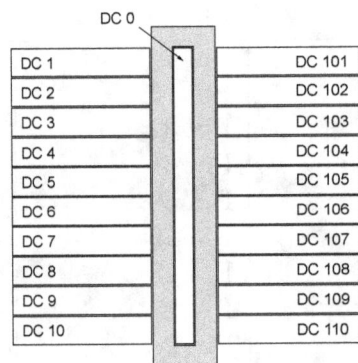

Figure 11.5: Electrode assignment of the planar linear trap

Electrode Assignment		Pin Assignment		DAC channel	
1	DC 1	Plug III	Pin 1	IC 9	Channel 1
	DC 101	Plug I	Pin 23	IC 7	Channel 0
2	DC 2	Plug III	Pin 14	IC 8	Channel 1
	DC 102	Plug I	Pin 22	IC 13	Channel 0
3	DC 3	Plug III	Pin 2	IC 3	Channel 1
	DC 103	Plug I	Pin 17	IC 5	Channel 1
4	DC 4	Plug III	Pin 15	IC 2	Channel 1
	DC 104	Plug I	Pin 4	IC 6	Channel 1
5	DC 5	Plug III	Pin 3	IC 1	Channel 1
	DC 105	Plug I	Pin 16	IC 7	Channel 1
6	DC 6	Plug III	Pin 16	IC 0	Channel 1
	DC 106	Plug I	Pin 3	IC 12	Channel 1
7	DC 7	Plug III	Pin 4	IC 11	Channel 0
	DC 107	Plug I	Pin 15	IC 13	Channel 1
8	DC 8	Plug III	Pin 17	IC 10	Channel 0
	DC 108	Plug I	Pin 2	IC 14	Channel 1
9	DC 9	Plug III	Pin 22	IC 9	Channel 0
	DC 109	Plug I	Pin 14	IC 15	Channel 1
10	DC 10	Plug III	Pin 23	IC 3	Channel 0
	DC 110	Plug I	Pin 1	IC 0	Channel 2
	DC 0	Plug III	Pin 24	IC 1	Channel 0
	Ground	Plug I	Pin 7, 20	Ground	
	Ground	Plug III	Pin 7, 19	Ground	

11.2.4 Planar microchip trap

The electrode assignment of this novel linear planar trap (Fig. 11.6) with an Y-shaped design shows a separated wiring of the electrodes for each segment pair: 57 dc electrodes are voltage controlled separately. The 5-segment loading zone DC1...DC5 and DC101...DC105 is connected by the 6-segment transfer zone DC6...DC11 and DC106...DC111 to the nodal point. The 7-segment storage zone DC19...DC25 and DC119...DC125 and the 7-segment processing zone DC12...DC18 and DC112...DC118 are located beyond that nodal point. The contact pads of the continuous center electrode DC200, the endcap electrodes DC70 and DC170, DC80 and DC180, DC90 and DC190 and the rf electrodes (dark gray) are located outside of the trap regions. The trap is embedded in a large ground area (light gray).

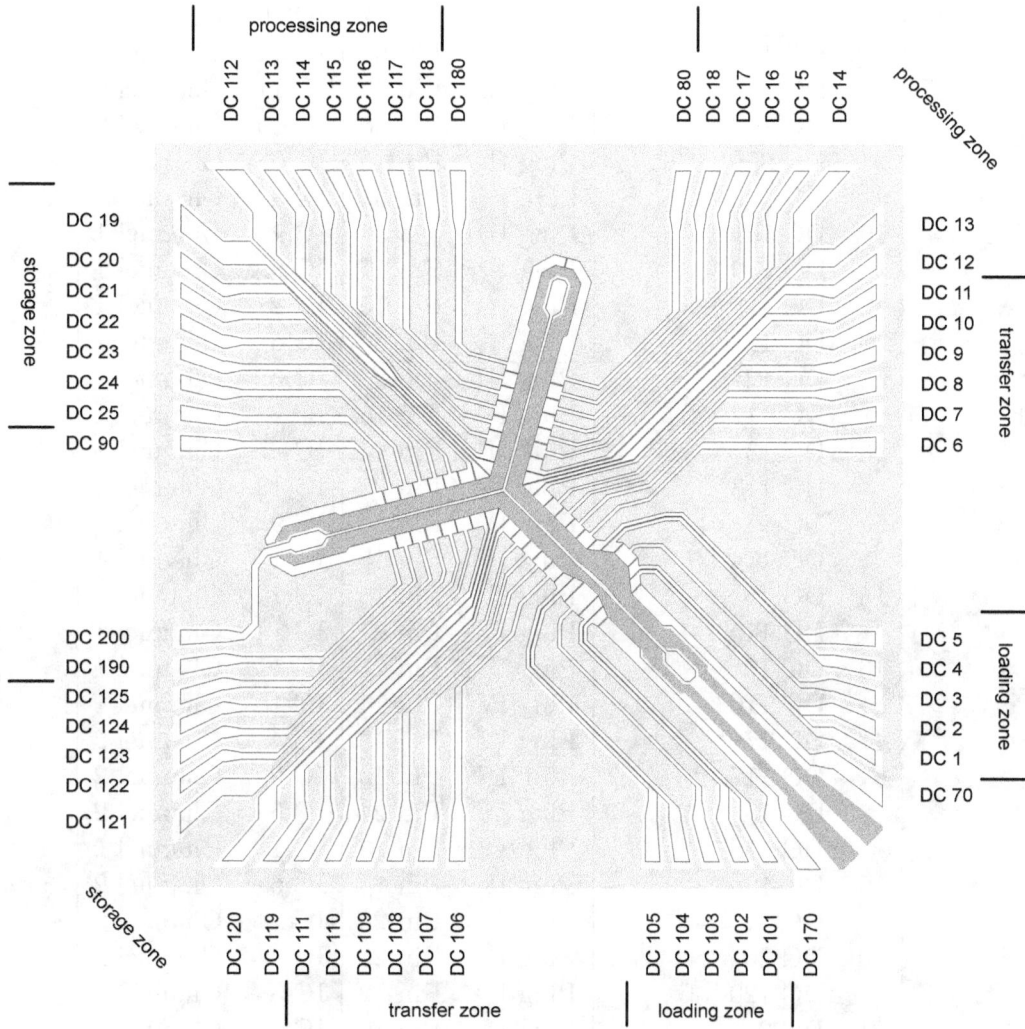

Figure 11.6: Electrode assignment of the planar microchip trap

Electrode Assignment		Pin Assignment		DAC channel	
1	DC 1	Plug III	Pin 12	IC 0	Channel 0
	DC 101	Plug I	Pin 12	IC 4	Channel 0
2	DC 2	Plug III	Pin 24	IC 1	Channel 0
	DC 102	Plug I	Pin 24	IC 5	Channel 0
3	DC 3	Plug III	Pin 11	IC 2	Channel 0
	DC 103	Plug I	Pin 11	IC 6	Channel 0
4	DC 4	Plug III	Pin 23	IC 3	Channel 0
	DC 104	Plug I	Pin 23	IC 7	Channel 0
5	DC 5	Plug III	Pin 10	IC 8	Channel 0
	DC 105	Plug I	Pin 10	IC 12	Channel 0
6	DC 6	Plug III	Pin 22	IC 9	Channel 0
	DC 106	Plug I	Pin 22	IC 13	Channel 0
7	DC 7	Plug III	Pin 17	IC 10	Channel 0
	DC 107	Plug I	Pin 9	IC 14	Channel 0
8	DC 8	Plug III	Pin 4	IC 11	Channel 0
	DC 108	Plug I	Pin 18	IC 15	Channel 0
9	DC 9	Plug III	Pin 16	IC 0	Channel 1
	DC 109	Plug I	Pin 5	IC 4	Channel 1
10	DC 10	Plug III	Pin 3	IC 1	Channel 1
	DC 110	Plug I	Pin 17	IC 5	Channel 1
11	DC 11	Plug III	Pin 15	IC 2	Channel 1
	DC 111	Plug I	Pin 4	IC 6	Channel 1
12	DC 12	Plug III	Pin 2	IC 3	Channel 1
	DC 112	Plug IV	Pin 16	IC 7	Channel 1
13	DC 13	Plug III	Pin 14	IC 8	Channel 1
	DC 113	Plug IV	Pin 3	IC 12	Channel 1
14	DC 14	Plug IV	Pin 1	IC 9	Channel 1
	DC 114	Plug IV	Pin 15	IC 13	Channel 1
15	DC 15	Plug IV	Pin 13	IC 10	Channel 1
	DC 115	Plug IV	Pin 2	IC 14	Channel 1
16	DC 16	Plug IV	Pin 25	IC 11	Channel 1
	DC 116	Plug IV	Pin 14	IC 15	Channel 1
17	DC 17	Plug IV	Pin 12	IC 4	Channel 2
	DC 117	Plug IV	Pin 1	IC 0	Channel 2
18	DC 18	Plug IV	Pin 24	IC 5	Channel 2
	DC 118	Plug IV	Pin 13	IC 1	Channel 2
19	DC 19	Plug II	Pin 11	IC 6	Channel 2
	DC 119	Plug I	Pin 25	IC 2	Channel 2
20	DC 20	Plug II	Pin 23	IC 7	Channel 2
	DC 120	Plug I	Pin 12	IC 3	Channel 2
21	DC 21	Plug II	Pin 10	IC 12	Channel 2
	DC 121	Plug II	Pin 24	IC 8	Channel 2

Electrode Assignment		Pin Assignment		DAC channel	
22	DC 22	Plug II	Pin 22	IC 13	Channel 2
	DC 122	Plug II	Pin 11	IC 9	Channel 2
23	DC 23	Plug II	Pin 9	IC 14	Channel 2
	DC 123	Plug II	Pin 23	IC 10	Channel 2
24	DC 24	Plug II	Pin 21	IC 15	Channel 2
	DC 124	Plug II	Pin 10	IC 11	Channel 2
25	DC 25	Plug II	Pin 5	IC 4	Channel 3
	DC 125	Plug II	Pin 22	IC 0	Channel 3
	DC 200	Plug II	Pin 5	IC 4	Channel 3
	DC 70	Plug III	Pin 25	Ground	
	DC 170	Plug I	Pin 25	Ground	
	DC 80	Plug IV	Pin 13	Ground	
	DC 180	Plug IV	Pin 13	Ground	
	DC 90	Plug II	Pin 1	Ground	
	DC 190	Plug II	Pin 1	Ground	
	Ground	Plug I	Pin 7, 20	Ground	
	Ground	Plug II	Pin 7, 20	Ground	
	Ground	Plug III	Pin 7, 19	Ground	
	Ground	Plug IV	Pin 7, 19	Ground	

The location of the contact pads provides an extended optical access. A cross-shaped area in the center tilted to the Y-shaped linear electrode geometry is leaved out from contact pads respectively bond wires. Therefore the loading region with the asymmetric electrode design is accessibly with lasers for Doppler cooling under an angle of 45° to the linear trap axis. The nodal point is centered on the microfabricated trap, linked continuously with the loading region regarding to the optical access. The bond pads for the dc electrodes are placed to a inboard position facilitating a revolving ground layer between all dc electrodes. The integration of vias for a multi-layer fabrication is avoided to minimize the error rate of the trap fabrication.

11.3   Rotational wafer coating

The fabrication of the multi-layer microtrap requires a double-sided closed gold coating for the two Al_2O_3 wafers carrying the trap electrodes. The deposition of the conductive material to a closed layer is crucial for the trap operation. Especially at the top layer the bond pads and the electrode surface generating the trapping potential are located on opposite sides, the electrical connection is guided along the head side of the finger electrodes. Besides the trapping potentials the conductive layer covers blank Al_2O_3 areas to prevent distortions of the electric field by local stray charges located at small areas of uncoated Al_2O_3.

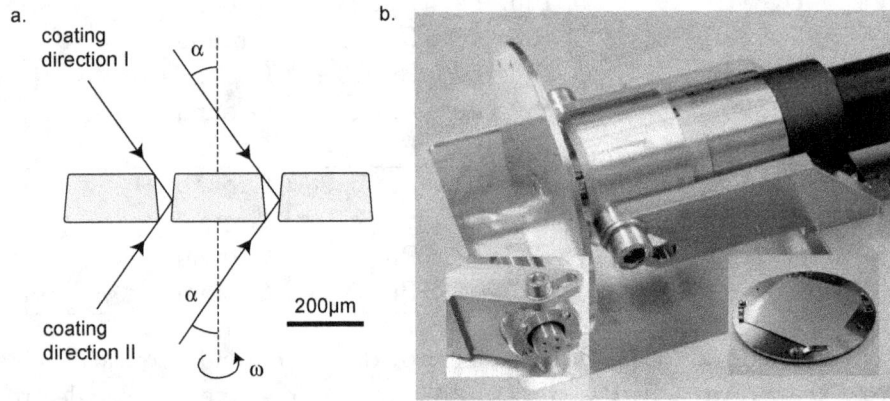

Figure 11.7: Conductive layer deposition: (a) Cross section of a single microtrap wafer in a double-sided deposition scheme (genuine dimensions). (b) Self-developed high-vacuum compatible single axis rotation stage with 75mm diameter rotary mount with variable inclination ($\alpha = 0°...90°$).

A continuous rotation during the electron beam physical vapor deposition (EBPVD) is required to cover the sidewalls of the laser cutted slits (Fig. 11.7a). Depending on the thickness of the wafer and the width of the laser cutted slits the inclination is adjusted. A rotation stage with continuously variable inclination was developed for the installation in the electron beam coating machine (Fig. 11.7b). Equipped with a high-vacuum compatible dc motor[2] the rotation stage allows a single-sided deposition of a 50mm squared Al_2O_3 wafer. The wafer is flipped after the first coating run, a second run is needed to close the conducting layer. The coating consists out of a 50nm thick titanium layer as adhesive layer, followed by a gold layer of 0.5μm thickness. The deposition rate was stabilized to 0.45nm/s, the rotation speed adjusted to 40rpm and the sample temperature limited to 120°C. Using a 50mm squared base Al_2O_3 wafer the production of four complete microchip traps in a single processing run was achieved.

---

[2]Faulhaber GmbH, Schönaich, Germany

a.
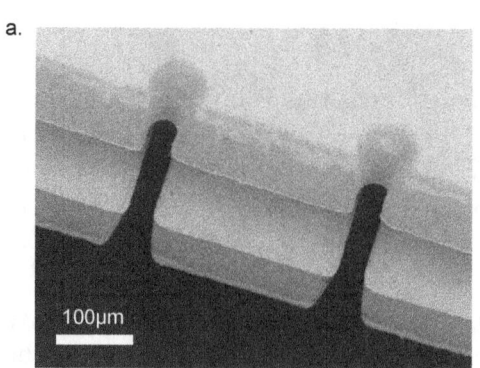

b.

Figure 11.8: Comparison by SEM of EBPVD layesr with and without sample rotation: Two prototypes of the two-layer microtrap are shown with a double-sided gold coating for each wafer. (a) Gold coating under normal incidence. The uncoated areas at the sidewalls are identified by the darker color. (b) Closed gold coating with deposition unter continuous rotation (rotation axis tilted by 35° to normal axis).

The coating quality is investigated with scanning electron microscopy (SEM). Different assembled microtrap prototypes are tested, the variation of the contrast at the sidewalls shows clearly the difference of the layer deposition under normal incidence (Fig. 11.8a) and with an inclination during continuous rotation of the sample (Fig. 11.8b). The material dependent visual change of the optical contrast is verified by spatial energy dispersive x-ray spectroscopy (EDX) used simultaneously during the characterization with SEM. The bright colored areas are gold coated, the rough dark regions are blank Al_2O_3. Based on the deposition rate the surface roughness of the gold coated wafer reflects the wafer surface quality directly.

11.4 Chip carrier socket for UHV

The experiments towards the development of a quantum computer show
the demands of standardized technical parts for electrical interconnect of
the ion trap to the laboratory electronics. This research work confirms
the applicability of leadless ceramic chip carriers (LCCC) for the supply of
multiple control voltages and even the radiofrequency with peak voltages of
several hundreds volts.

The leadless chip carriers (LCC) are privileged towards pin grid arrays
(PGA): Their straddled rectilinear contact lines inside the ceramics of the
carrier minimizes rf pickup and crossed lines are avoided compared to PGA
carriers. The outer contact pads are spaced sufficiently for an easy access
with standard single-layer printed circuit boards (PCB). This allows the
integration of low-pass filters close to the chip carrier, which requires multi-
layer PCB. The measured magnetic stray fields from the PGA pins are orders
of magnitudes higher at the position of the ions than the residual fields from
the other PGA/LCC components. Additionally the surface mounting of the
chip carrier (LCC) prevent virtual leaks in contrast to a standard pin/socket
connection (PGA).

a.
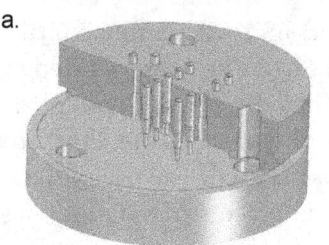
b.

Figure 11.9: Demonstration model (a) for electrical contacting of the
microtrap chip carrier in the UHV using spring contact probe technology.
The contact spacing is 1.27mm, the utility model (b) is used for testing
the contact pressure after vacuum beakout with a maximum temperature
of 150°C and measuring the outgassing properties.

The ion traps used in this work are installed in a LCC, which is soldered
to the PCB directly. Providing a fast exchange, an UHV compatible scalable chip carrier socket was designed: The surface mounted LCC is placed
on spring probes for electrical contact by a metallic clamp. The bottom
wings of the outer LCC contacts serve as electrical interconnects. The UHV
compatible spring probes 100881[3] are vented to avoid virtual leaks and with-
stand bakeout temperatures of 150°C. Even the radiofrequency trap drive
up to currents of 6A and several 10MHz of frequency can be supplied by
the 0.75mm diameter and 5.9mm long spring probes.

---

[3]Interconnect Devices Inc., Kansas, USA

The design is simplified to a single probe mount made out of machinable glass ceramic Macor[4], wherein the spring probes stick and are pressed to the PCB. A prototype with a pitch identical to contact pad distance of the LCC was fabricated and tested (Fig. 11.9).

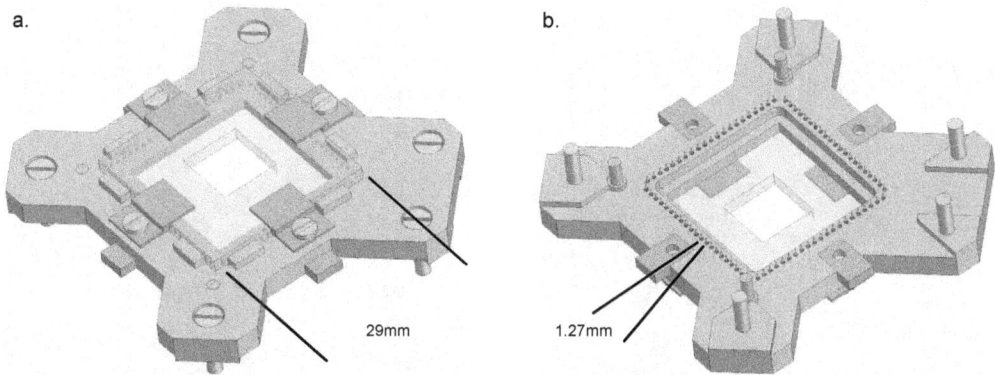

Figure 11.10: Design model (genuine dimensions) of a UHV compatible ceramic chip carrier socket with a variable pitch, which provides mounting of LCC type components. The front side (a) shows the metal clamps (gold) pressing the carrier (transparent) on the spring probes. The macor mount (blue) is attached to the PCB by screws. At the back side (b) the arrangement of the spring probes is shown.

A realistic design for the LCC used in the experiments (Fig. 11.10) is deduced from the demonstration model (Fig. 11.9). All contact pads of the chip carrier can be controlled separately by 84 spring probes. The open coverage type is required for optimized optical access by a large numerical aperture. Geometric laser configurations for multi-layer microtraps perpendicular to the chip carrier and planar traps parallel to the chip carrier are supported. The chip carrier socket provides the mounting on a low-pass filter board, then the filter components are located close to the chip carrier for efficient rejection of rf noise. Moreover the grid array of the spring probe arrangement can be adapted easily to other geometrical requirements (i.e. BGA). For the design of the UHV compatible chip carrier socket the patent is pending (Nr. DE102006023158A1, 2006).

---

[4]Corning Inc., New York, USA.

Bibliography

[Ahn00]   J. Ahn, T. C. Weinacht, and P. H. Bucksbaum, *Information Storage and Retrieval Through Quantum Phase*, Science 287, 463 (2000).

[Alh95]   R. Alheit, C. Hennig, R. Morgenstern, F. Vedel, and G. Werth, *Observation of instabilities in a Paul trap with higher-order anharmonicities*, Appl. Phys. B 61, 277 (1995).

[Alh96]   R. Alheit, S. Kleineidam, F. Vedel, M. Vedel, and G. Werth, *Higher and non-linear resonances in a Paul trap*, Int. J. Mass. Spec. 154, 155 (1996).

[Aln08]   J. Alnis, A. Matveev, N. Kolachevsky, T. Udem, and T. W. Hänsch, *Subhertz linewidth diode lasers by stabilization to vibrationally and thermally compensated ultralow-expansion glass Fabry-Perot cavities*, Phys. Rev. A 77, 053809 (2008).

[Ami08]   J. M. Amini, J. Britton, D. Leibfried, and D. J. Wineland, *Microfabricated chip traps for ions*, arXiv:0812.3907 (2008).

[And06]   A. Andre, D. DeMille, J. M. Doyle, M. D. Lukin, S. E. Maxwell, P. Rabl, R. J. Schoelkopf, and P. Zoller, *A coherent all-electrical interface between polar molecules and mesoscopic superconducting resonators*, Nature 2, 636 (2006).

[Ant08]   P. B. Antohi, D. Schuster, G. M. Akselrod, J. Labaziewicz, Y. Ge, Z. Lin, W. S. Bakr, and I. L. Chuang, *Cryogenic ion trapping systems with surface-electrode traps*, arXiv:0807.4763 (2008).

[Bab02]   T. Baba and I. Waki, *Sympathetic cooling rate of gas-phase ions in a radio-frequency-quadrupole ion trap*, Appl. Phys. B 74, 375 (2002).

[Bad00]   E. R. Badman and R. G. Cooks, *A parallel miniature cylindrical ion trap array*, Anal. Chem. 72, 3291 (2000).

[Bar03]   M. Barrett, B. L. DeMarco, T. Schätz, V. Meyer, D. Leibfried, J. Britton, J. Chiaverini, W. M. Itano, B. M. Jelenkovic, J. D. Jost, C. Langer, T. Rosenband, and D. J. Wineland, *Sympathetic cooling of* $^9Be^+$ *and* $^{24}Mg^+$ *for quantum logic*, Phys. Rev. A 68, 042302 (2003).

[Bar04]   M. D. Barrett, J. Chiaverini, T. Schätz, J. Britton, W. M. Itano, J. D. Jost, E. Knill, C. Langer, D. Leibfried, R. Ozeri and D. J. Wineland, *Deterministic quantum teleportation of atomic qubits*, Nature 429, 737 (2004).

[Bec01]   T. Becker, J. v. Zanthier, A. Nevsky, C. Schwedes, M. N. Skvortsov, H. Walther, and E. Peik, *High-resolution spectroscopy of a single* In^+ *ion: Progress towards an optical frequency standard*, Phys. Rev. A 63, 051802 (2001).

[Ben08a] J. Benhelm, G. Kirchmair, C. F. Roos, and R. Blatt, *Towards fault-tolerant quantum computing with trapped ions*, Nature Physics 4, 463 (2008).

[Ben08b] J. Benhelm, *Precision spectroscopy and quantum information processing with trapped calcium ions*, Doctoral thesis, University of Innsbruck (2008).

[Ben80]   P. Benioff, *The computer as a physical system: A microscopic quantum mechanical Hamiltonian model of computers as represented by Turing machines*, J. Stat. Phys. 22, 563 (1980).

[Ben82]   P. Benioff, *Quantum mechanical models of Turing machines that dissipate no energy*, Phys. Rev. Lett. 48, 1581 (1982).

[Ber86]   J. C. Bergquist, R. Hulet, W. M. Itano, and D. J. Wineland, *Observation of quantum jumps in a single atom*, Phys. Rev. Lett. 57, 1699 (1986).

[Ber98a] D. J. Berkeland, J. D. Miller, J. C. Bergquist, W. M. Itano, and D. J. Wineland, *Laser-cooled mercury ion frequency standard*, Phys. Rev. Lett. 80, 2089 (1998).

[Ber98b] D. J. Berkeland, J. D. Miller, J. C. Bergquist, W. M. Itano, and D. J. Wineland, *Minimization of the ion micromotion in a Paul trap*, J. Appl. Phys. 83, 5025 (1998).

[Ber99]   J. E. Bernard, A. A. Madej, L. Marmet, B. G. Whitford, K. J. Siemsen, and S. Cundy, *Cs-based frequency measurement of a single, trapped ion transition in the visible region of the spectrum*, Phys. Rev. Lett. 82, 3228 (1999).

[Biz03]   S. Bize, S. A. Diddams, U. Tanaka, C. E. Tanner, W. H. Oskay, R. E. Drullinger, T. E. Parker, T. P. Heavner, S. R. Jefferts, L. Hollberg, W. M. Itano, and J. C. Bergquist, *Testing the stability of fundamental constants with the ^{199}Hg$^+$ single-ion optical clock*, Phys. Rev. Lett. 90, 150802 (2003).

[Bly05]   P. Blythe, B. Roth, U. Fröhlich, H. Wenz, and S. Schiller, *Production of ultracold trapped molecular hydrogen ions*, Phys. Rev. Lett. 95, 183002 (2005).

[Boc04]   A. Boca, R. Miller, K. M. Birnbaum, A. D. Boozer, J. McKeever, and H. J. Kimble, *Observation of the vacuum Rabi spectrum for one trapped atom*, Phys. Rev. Lett. 93, 233603 (2004).

[Bol96]   J. J. Bollinger, W. M. Itano, and D. J. Wineland, *Optimal frequency measurements with maximally correllated states*, Phys. Rev. A 54, 4649 (1996).

[Bow99]   P. Bowe, L. Hornekaer, C. Brodersen, M. Drewsen, J. S. Hangst, and J. P. Schiffer, *Sympathetic crystallization of trapped ions*, Phys. Rev. Lett. 82, 2071 (1999).

[Boy06]   M. M. Boyd, T. Zelevinsky, A. D. Ludlow, S. M. Foreman, S. Blatt, T. Ido, and J. Ye, *Optical atomic coherence at the 1-second time scale*, Science 314, 1430 (2006).

[Boy07]   M. M. Boyd, A. D. Ludlow, S. Blatt, S. M. Foreman, T. Ido, T. Zelevinsky, and J. Ye, ^{87}Sr *Lattice clock with inaccuracy below* 10^{-15}, Phys. Rev. Lett. 98, 083002 (2007).

[Bri05]   K.-A. Brickman, P. C. Haljan, P. J. Lee, M. Acton, L. Deslauriers, and C. Monroe, *Implementation of Grovers quantum search algorithm in a scalable system*, Phys. Rev. A 72, 050306 (2005).

[Bri06]   J. Britton, D. Leibfried, J. Beall, R. B. Blakestad, J. J. Bollinger, J. Chiaverini, R. J. Epstein, J. D. Jost, D. Kielpinski, C. Langer, R. Ozeri, R. Reichle, S. Seidelin, N. Shiga, J. H. Wesenberg, and D. J. Wineland, *A microfabricated surface-electrode ion trap in silicon*, arXiv:quant-ph/0605170 (2006).

[Bro07]   K. R. Brown, R. J. Clark, J. Labaziewicz, P. Richerme, D. R. Leibrandt, and I. L. Chuang, *Loading and characterization of a printed-circuit-board atomic ion trap*, Phys. Rev. A 75, 015401 (2007).

[Bus06]   P. Bushev, D. Rotter, A. Wilson, F. Dubin, C. Becher, J. Eschner, R. Blatt, V. Steixner, P. Rabl, and P. Zoller, *Feedback cooling of a single trapped ion*, Phys. Rev. Lett. 96, 043003 (2006).

[Cal04]   T. Calarco, U. Dorner, P. S. Julienne, C. J. Williams, and
           P. Zoller, *Quantum computations with atoms in optical lattices:
           Marker qubits and molecular interactions*, Phys. Rev. A 70,
           012306 (2004).

[Chi04]   J. Chiaverini, D. Leibfried, T. Schaetz, M. D. Barrett,
           R. B. Blakestad, J. Britton, W. M. Itano, J. D. Jost, E. Knill,
           C. Langer, R. Ozeri and D. J. Wineland, *Realization of quantum
           error correction*, Nature 432, 602 (2004).

[Chi05]   J. Chiaverini, R. B. Blakestad, J. Britton, J. D. Jost, C. Langer,
           D. Leibfried, R. Ozeri, and D. J. Wineland, *Surface-electrode
           architecture for ion-trap quantum information processing*, Quan-
           tum Information and Computation, Vol. 5, 419 (2005).

[Chu]   I. L. Chuang, private communication.

[Chu98]   I. L. Chuang, N. Gershenfeld, and M. Kubinec, *Experimental
           implementation of fast quantum searching*, Phys. Rev. Lett. 80,
           3408 (1998).

[Cir00]   J. I. Cirac and P. Zoller, *A scalable quantum computer with ion
           in an array of microtraps*, Nature 404, 579 (2000).

[Cir92]   J. I. Cirac, R. Blatt, P. Zoller and W. D. Phillips, *Laser cooling
           of trapped ions in a standing wave*, Phys. Rev. A 46, 2668 (1992).

[Cir95]   J. I. Cirac and P. Zoller, *Quantum computations with cold trapped
           ions*, Phys. Rev. Lett. 74, 4091 (1995).

[Coh99]   C. Cohen-Tannoudji, B. Diu, and F. Laloe, *Quantenmechanik
           Bd. 1+2*, Berlin (1999).

[Col07]   Y. Colombe, T. Steinmetz, G. Dubois, F. Linke, D. Hunger, and
           J. Reichel, *Strong atom-field coupling for Bose-Einstein conden-
           sates in an optical cavity on a chip*, Nature 450, 272 (2007).

[Deh67a]   H. G. Dehmelt, *Radiofrequency spectroscopy of stored ions I:
           Storage*, Adv. At. Mol. Phys. 3, 53 (1967).

[Deh67b]   H. G. Dehmelt, *Radiofrequency spectroscopy of stored ions II:
           Spectroscopy*, Adv. At. Mol. Phys. 5, 109 (1967).

[Dem02]   D. DeMille, *Quantum computation with trapped polar molecules.*
           Phys. Rev. Lett. 88, 067901 (2002).

[Des04]   L. Deslauriers, P. C. Haljan, P. J. Lee, K.-A. Brickman, B. B. Bli-
           nov, M. J. Madsen, and C. Monroe, *Zero-point cooling and low
           heating of trapped $^{111}Cd^+$ ions*, Phys. Rev. A 70, 043408 (2004).

[Des06]   L. Deslauriers, S. Olmschenk, D. Stick, W. K. Hensinger, J. Sterk, and C. Monroe, *Scaling and suppression of anomalous heating in ion traps*, Phys. Rev. Lett. 97, 103007 (2006).

[Deu85]   D. Deutsch. *Quantum theory, the Church-Turing principle and the universal quantum computer*. Proc. R. Soc. London A 400, 96 (1985).

[Deu89]   D. Deutsch. *Quantum computational networks*. Proc. R. Soc. London Ser. A 425, 73 (1989).

[Die87]   F. Diedrich, E. Peik, J. M. Chen, W. Quint, and H. Walther, *Observation of a phase transition of stored laser-cooled ions*, Phys. Rev. Lett. 59, 2931 (1987).

[Die89]   F. Diedrich, J. C. Bergquist, W. M. Itano, and D. J. Wineland, *Laser cooling to the zero-point energy of motion*, Phys. Rev. Lett. 62, 403 (1989).

[Div00]   D. P. DiVincenzo, *The physical implementation of quantum computation*, Fortschr. Phys. 48, 771 (2000).

[Dor05]   U. Dorner, T. Calarco, P. Zoller, A. Browaeys, and P. Grangier, *Quantum logic via optimal control in holographic dipole traps*, J. Opt. B. 7, S341 (2005).

[Dre00]   M. Drewsen and A. Broner, *Harmonic linear Paul trap: Stability diagram and effective potentials*, Phys. Rev. A, 62, 045401 (2000).

[Dre04]   M. Drewsen, A. Mortensen, R. Martinussen, P. Staanum, and J. L. Soerensen, *Nondestructive identification of cold and extremely localized single molecular ions*, Phys. Rev. Lett. 93, 243201 (2004).

[Dre98]   M. Drewsen, C. Brodersen, L. Hornekaer, J. S. Hangst, and J. P. Schiffer, *Large ion crystals in a linear Paul trap*, Phys. Rev. Lett. 81, 2878 (1998) and private communication: The trap drive U_{rf} is applied also to the dc electrodes with opposite phase.

[Ebl07]   J. Eble and F. Schmidt-Kaler, *Optimization of frequency modulation transfer spectroscopy on the calcium 4^1S_0 to 4^1P_1 transition*, Appl. Phys. B 88, 563 (2007).

[Esc03]   J. Eschner, G. Morigi, F. Schmidt-Kaler, and R. Blatt, *Laser cooling of trapped ions*, J. Opt. Soc. Am. B 20, 1003 (2003).

[Fey82]   R. Feynman, *Simulating physics with computers*, Int. J. Theor. Phys. 21, 467 (1982).

[For07]   T. M. Fortier, N. Ashby, J. C. Bergquist, M. J. Delaney, S. A. Diddams, T. P. Heavner, L. Hollberg, W. M. Itano, S. R. Jefferts, K. Kim, F. Levi, L. Lorini, W. H. Oskay, T. E. Parker, J. Shirley, and J. E. Stalnaker, *Precision atomic spectroscopy for improved limits on variation of the fine structure constant and local position invariance*, Phys. Rev. Lett. 98, 070801 (2007).

[Fre99]   C. B. Freidhoff, R. M. Young, S. Sriram, T. T. Braggins, T. W. O'Keefe, J. D. Adam, H. C. Nathanson, R. R. A. Syms, T. J. Tate, M. M. Ahmad, S. Taylor, and J. Tunstall, *Chemical sensing using nonoptical microelectromechanical systems*, J. Vac. Sci. Technol. A 17, 2300 (1999).

[Fri08]   A. Friedenauer, H. Schmitz, J. Glueckert, D. Porras and T. Schätz, *Simulating a quantum magnet*, Nature Physics 4, 757 (2008).

[Gar03]   J. J. Garcia-Ripoll, P. Zoller, and J. I. Cirac, *Speed optimized two-qubit gates with laser coherent control techniques for ion trap quantum computing*, Phys. Rev. Lett. 91, 157901 (2003).

[Gee05]   M. Geear, R. R. A. Syms, S. Wright, and A. S. Holmes, *Monolithic MEMS quadrupole mass spectrometers by deep silicon etching*, J. Microelectromechanical Systems 14, 1156 (2005).

[Gol89]   G. H. Golub and C. F. Van Loan, *Matrix computations*, 2nd ed., John Hopkins University press, Baltimore, 1989.

[Gor05]   J. Gorman, D. G. Hasko, and D. A. Williams, *Charge-qubit operation of an isolated double quantum dot*, Phys. Rev. Lett. 95, 090502 (2005).

[Gos95]   P. K. Gosh, *Ion traps*, Oxford (1995).

[Gro97]   L. K. Grover, *Quantum mechanics helps in searching for a needle in a haystack*, Phys. Rev. Lett. 79, 325 (1997).

[Gul01a]   S. Gulde, D. Rotter, P. Barton, F. Schmidt-Kaler, R. Blatt, and W. Hogervorst, *Simple and efficient photo-ionization loading of ions for precision ion-trapping experiments.* Appl. Phys. B, 73, 861 (2001).

[Gul01b]   M. S. Gulley, A. G. White, D. F. V. James, *A Raman approach to quantum logic in Calcium-like ions*, arXiv:quantph/0112117 (2001).

[Gul03]   S. Gulde, M. Riebe, G. T. Lancaster, C. Becher, J. Eschner, H. Häffner, F. Schmidt-Kaler, I. L. Chuang, and Rainer Blatt, *Implementation of the Deutsch-Jozsa algorithm on an ion-trap quantum computer*, Nature 421, 48 (2003).

[Hae05]   H. Häffner, W. Hänsel, C. F. Roos, J. Benhelm, D. Chek-al-kar, M. Chwalla, T. Körber, U. D. Rapol, M. Riebe, P. O. Schmidt, C. Becher, O. Gühne, W. Dür, and R. Blatt, *Scalable multi-particle entanglement of trapped ions*, Nature 438, 643 (2005).

[Hae75]   T. Hänsch and A. Schawlow, *Cooling of gases by laser radiation*, Opt. Commun. 13, 68 (1975).

[Hal06]   J. L. Hall, *Defining and measuring optical frequencies: the optical clock opportunity - and more*, ChemPhysChem 7, 2242 (2006).

[Hen06]   W. K. Hensinger, S. Olmschenk, D. Stick, D. Hucul, M. Yeo, M. Acton, L. Deslauriers, C. Monroe, and J. Rabchuk, *T-junction ion trap array for two-dimensional ion shuttling, storage, and manipulation*, Appl. Phys. Lett. 88, 034101 (2006).

[Hoj08]   K. Hojbjerre, D. Offenberg, C. Z. Bisgaard, H. Stapelfeldt, P. Staanum, A. Mortensen and M. Drewsen, *Consecutive photodissociation of a single complex molecular ion*, Phys. Rev. A 77, 030702 (2008).

[Hol04]   L. C. Hollenberg, A. S. Dzurak, C. Wellard, A. R. Hamilton, D. J. Reilly, G. J. Milburn, and R. G. Clark, *Charge-based quantum computing using single donors in semiconductors*, Phys. Rev. B 69, 113301 (2004).

[Hom06a]   J. Home, *Entanglement of two trapped-ion spin qubits*, Doctoral thesis, Oxford (2006).

[Hom06b]   J. P. Home and A. M. Steane, *Electrode configurations for fast separation of trapped ions*, Quant. Inf. Comput. 6, 289 (2006).

[Hou08]   M. G. House, *Analytic model for electrostatic fields in surface-electrode ion traps*, Phys. Rev. A 78, 033402 (2008).

[Hub08]   G. Huber, T. Deuschle, W. Schnitzler, R. Reichle, K. Singer, and F. Schmidt-Kaler, *Transport of ions in a segmented linear Paul trap in printed-circuit-board technology*, New J. Phys. 10, 013004 (2008).

[Ita82]   W. M. Itano and D. J. Wineland, *Laser cooling of ions stored in harmonic and Penning traps*, Phys. Rev. A 25, 35 (1982).

[Ita95]   W. M. Itano, J. C. Bergquist, J. J. Bollinger, and D. J. Wineland, *Cooling methods in ion traps*, Physica Scripta T59, 106 (1995).

[Jam98]   D. F. V. James, *Quantum dynamics of cold trapped ions with application to quantum computation*, Appl. Phys. B 66, 181 (1998).

[Jef02]   S. R. Jefferts, J. Shirley, T. E. Parker, T. P. Heaver, D. M. Meekhof, C. Nelson, F. Levi, G. Costanzo, A. De Marchi, R. Drullinger, L. Hollberg, W. D. Lee, and F. L. Walls, *Accuracy evaluation of NIST-F1*, Metrologia 39, 321 (2002).

[Jon98]   J. A. Jones, M. Mosca, and R. H. Hansen, *Implementation of a quantum search algorithm on a quantum computer*, Nature 393, 344 (1998).

[Kie92]   D. Kielpinski, C. Monroe and D. J. Wineland, *Architecture for a large-scale ion-trap quantum computer*, Nature 417, 709 (2002).

[Kin98]   B. E. King, C. S. Wood, C. J. Myatt, Q. A. Turchette, D. Leibfried, W. M. Itano, C. Monroe, and D. J. Wineland, *Cooling the collective motion of trapped ions to initialize a quantum register*, Phys. Rev. Lett. 81, 1525 (1998).

[Kir04]   D. Kirk, *Optimal control theory - an introduction*, Dover Publications, New York (2004).

[Kni01]   E. Knill, R. Laflamme, and G. J. Milburn, *A scheme for efficient quantum computation with linear optics*, Nature 409, 46 (2001).

[Kni05]   E. Knill, *Quantum computing with realistic noisy devices*, Nature 434, 39 (2005).

[Koe07]   J. C. J. Koelemeij, B. Roth, A. Wicht, I. Ernsting, and S. Schiller, *Vibrational spectroscopy of* HD^+ *with 2-ppb accuracy*, Phys. Rev. Lett. 98, 173002 (2007).

[Kok07]   P. Kok, W. J. Munro, K. Nemoto, T. C. Ralph, J. P. Dowling, and G. J. Milburn, *Linear optical quantum computing with photonic qubits*, Rev. Mod. Phys. 79, 135 (2007).

[Lab08a]   J. Labaziewicz, Y. Ge, P. Antohi, D. Leibrandt, K. R. Brown, and I. L. Chuang, *Suppression of heating rates in cryogenic surface-electrode ion traps*, Phys. Rev. Lett. 100, 013001 (2008).

[Lab08b]   J. Labaziewicz, Y. Ge, D. Leibrandt, S. X. Wang, R. Shewmon, and I. L. Chuang, *Temperature dependence of electric field noise above gold surfaces*, arXiv:0804.2665 (2008).

[Lam07]   L. Lamata, J. Leon, T. Schätz, and E. Solano, *Dirac equation and quantum relativistic effects in a single trapped ion*, Phys. Rev. Lett. 98, 254005 (2007).

[Lan05]   C. Langer, R. Ozeri, J. D. Jost, J. Chiaverini, B. DeMarco, A. Ben-Kish, R. B. Blakestad, J. Britton, D. B. Hume, W. M. Itano, D. Leibfried, R. Reichle, T. Rosenband, T. Schaetz, P. O. Schmidt, and D. J. Wineland, *Long-lived qubit memory using atomic ions*, Phys. Rev. Lett. 95, 060502 (2005).

[Lei03a]   D. Leibfried, R. Blatt, C. Monroe, and D. Wineland, *Quantum dynamics of single trapped ions*, Rev. Mod. Phys. 75, 281 (2003).

[Lei03b]   D. Leibfried, B. DeMarco, V. Meyer, D. Lucas, M. Barrett, J. Britton, W. M. Itano, B. Jelenkovic, C. Langer, T. Rosenband, and D. J. Wineland. *Experimental demonstration of a robust, high-fidelity geometric two ion-qubit phase gate*. Nature 422, 412 (2003).

[Lei04]   D. Leibfried, M. D. Barrett, T. Schaetz, J. Britton, J. Chiaverini, W. M. Itano, J. D. Jost, C. Langer, and D. J. Wineland, *Toward Heisenberg-limited spectroscopy with multiparticle entangled states*, Science 304, 1476 (2004).

[Lei05]   D. Leibfried, E. Knill, S. Seidelin, J. Britton, R. B. Blakestad, J. Chiaverini, D. B. Hume, W. M. Itano, J. D. Jost, C. Langer, R. Ozeri, R. Reichle, and D. J. Wineland, *Creation of a six-atom 'Schrödinger cat' state*, Nature 438, 639 (2005).

[Lei07]   D. R. Leibrandt, R. J. Clark, J. Labaziewicz, P. Antohi, W. Bakr, K. R. Brown, and I. L. Chuang, *Laser ablation loading of a surface-electrode ion trap*, Phys. Rev. A 76, 055403 (2007).

[Lin98]   N. Linden, H. Barjat, and R. Freeman, *An implementation of the Deutsch-Jozsa algorithm on a three-qubit NMR quantum computer*, Chem. Phys. Lett. 296, 61 (1998).

[Llo95]   S. Lloyd, *Almost any quantum logic gate is universal*, Phys. Rev. Lett. 75, 346 (1995).

[Lud07]   A. D. Ludlow, X. Huang, M. Notcutt, T. Zanon-Willette, S. M. Foreman, M. M. Boyd, S. Blatt, and J. Ye, *Compact, thermal-noise-limited optical cavity for diode laser stabilization at $1 \cdot 10^{-15}$*, Opt. Lett. 32, 641 (2007).

[Mac59]   W. W. MacAlpine and R. O. Schildknecht, *Coaxial resonators with helical inner conductor*, Proc. of the IRE 47, 2099 (1959).

[Mar04]   H. S. Margolis, G. P. Barwood, G. Huang, H. A. Klein, S. N. Lea, K. Szymaniec, and P. Gill, *Hertz-level measurement of the optical clock frequency in a single $^{88}Sr^+$ ion*, Science 306, 1355 (2004).

[Mar94]   I. Marzoli, J. I. Cirac, R. Blatt, and P. Zoller, *Laser cooling of trapped three-level ions: Designing two-level systems for sideband cooling*, Phys. Rev. A 49, 2771 (1994).

[Mau05]   P. Maunz, T. Puppe, I. Schuster, N. Syassen, P. W. H. Pinkse, and G. Rempe, *Normal-mode spectroscopy of a single-bound-atom cavity system*, Phys. Rev. Lett. 94, 033002 (2005).

[Mcd05]   R. McDermott, R. W. Simmonds, M. Steffen, K. B. Cooper, K. Cicak, K. D. Osborn, S. Oh, D. P. Pappas, and J. M. Martinis, *Simultaneous state measurement of coupled Josephson phase qubits*, Science 307, 1299 (2005).

[Moe07]   D. L. Moehring, P. Maunz, S. Olmschenk, K. C. Younge, D. N. Matsukevich, L.-M. Duan and C. Monroe, *Entanglement of single-atom quantum bits at a distance*, Nature 449, 68 (2007).

[Mon95a] C. Monroe, D. M. Meekhof, B. E. King, S. R. Jefferts, W. M. Itano, and D. J. Wineland, *Resolved-sideband Raman cooling of a bound atom to the 3D zero-point energy*, Phys. Rev. Lett. 75, 4011 (1995).

[Mon95b] C. Monroe, D. M. Meekhof, B. E. King, W. M. Itano, and D. J. Wineland, *Demonstration of a fundamental quantum logic gate*, Phys. Rev. Lett. 75, 4714 (1995).

[Mor01]   G. Morigi and J. Eschner, *Doppler cooling of a Coulomb crystal*, Phys. Rev. A 64, 063407 (2001).

[Mor06]   A. Mortensen, E. Nielsen, T. Matthey, and M. Drewsen, *Observation of three-dimensional long-range order in small ion coulomb crystals in an rf trap*, Phys. Rev. Lett. 96, 103001 (2006).

[Mor97]   G. Morigi, J. I. Cirac, M. Lewenstein, and P. Zoller, *Ground state laser cooling beyond the Lamb-Dicke limit*, Europhys. Lett. 23, 1 (1997).

[Nag86]   W. Nagourney, J. Sandberg, and H. Dehmelt, *Shelved optical electron amplifier: observation of quantum jumps*, Phys. Rev. Lett. 56, 2797 (1986).

[Neu78]   W. Neuhauser, M. Hohenstatt, P. Toschek, and H. Dehmelt, *Optical-sideband cooling of visible atom cloud confined in a parabolic well*, Phys. Rev. Lett. 41, 223 (1978).

[Neu80]   W. Neuhauser, M. Hohenstatt, P. E. Toschek, and H. Dehmelt, *Localized visible* Ba^+ *mono-ion oscillator*, Phys. Rev. A 22, 1137 (1980).

[Nie00]   M. A. Nielsen and I. L. Chuang, *Quantum computation and quantum information*. Cambridge Univ. Press, Cambridge (2000).

[Not05]   M. Notcutt, L.-S. Ma, J. Ye, and J. L. Hall, *Simple and compact 1-Hz laser system via an improved mounting configuration of a reference cavity*, Opt. Lett., 30, 1815 (2005).

[Off08]   D. Offenberg, C. B. Zhang, Ch. Wellers, B. Roth, and S. Schiller, *Translational cooling and storage of protonated proteins in an ion trap at subkelvin temperatures*, Phys. Rev. A 78, 061401 (2008).

[Osk06]   W. H. Oskay, S. A. Diddams, E. A. Donley, T. M. Fortier, T. P. Heavner, L. Hollberg, W. M. Itano, S. R. Jefferts, M. J. Delaney, K. Kim, F. Levi, T. E. Parker, and J. C. Bergquist, *Single-atom optical clock with high accuracy*, Phys. Rev. Lett. 97, 020801 (2006).

[Oze07]   R. Ozeri, W. M. Itano, R. B. Blakestad, J. Britton, J. Chiaverini, J. D. Jost, C. Langer, D. Leibfried, R. Reichle, S. Seidelin, J. H. Wesenberg, and D. J. Wineland, *Errors in trapped-ion quantum gates due to spontaneous photon scattering*, Phys. Rev. A 75, 042329 (2007).

[Pau53]   W. Paul, and H. Steinwedel, *Ein neues Massenspektrometer ohne Magnetfeld*, Z. Naturforsch. 8a, 448 (1953).

[Pau55]   W. Paul, and M. Raether, *Das elektrische Massenfilter*, Z. Physik 140, 262 (1955).

[Pau90]   W. Paul, *Electromagnetic traps for charged and neutral particles*, Rev. Mod. Phys. 62, 531 (1990).

[Pea06]   C. E. Pearson, D. R. Leibrandt, W. S. Bakr, W. J. Mallard, K. R. Brown, and I. L. Chuang, *Experimental investigation of planar ion traps*, Phys. Rev. A 73, 032307 (2006).

[Pei04]   E. Peik, B. Lipphardt, H. Schnatz, T. Schneider, C. Tamm, and S. G. Karshenboim, *Limit on the present temporal variation of the fine structure constant*, Phys. Rev. Lett. 93, 170801 (2004).

[Pei99]   E. Peik, J. Abel, T. Becker, J. von Zanthier, and H. Walther, *Sideband cooling of ions in radio-frequency traps*, Phys. Rev. A 60, 439 (1999).

[Por04a] D. Porras and J. I. Cirac, *Effective quantum spin systems with trapped ions*, Phys. Rev. Lett. 92, 207901 (2004).

[Por04b] D. Porras and J. I. Cirac, *Bose-Einstein condensation and strong-correlation behavior of phonons in ion traps*, Phys. Rev. Lett. 93, 263602 (2004).

[Pre92]   H. M. Presby and C. A. Edwards, *Near 100% efficient fibre microlenses*, Electronics Letters, 28, 582 (1992).

[Rai92]   M. G. Raizen, J. M. Gilligan, J. C. Bergquist, W. M. Itano, and D. J. Wineland, *Ionic crystals in a linear Paul trap*, Phys. Rev. A 45, 6493 (1992).

[Ram05]   C. Ramanathan, N. Boulant, Z. Chen, D. G. Cory, I. Chuang and M. Steffen, *NMR quantum information processing*, Quant. Inf. Proc. 3, 15 (2005).

[Rau07]   R. Raussendorf and J. Harrington, *Fault-tolerant quantum computation with high treshold in two dimensions*, Phys. Rev. Lett. 98, 190504 (2007).

[Rei06a] R. Reichle, D. Leibfried, E. Knill, J. Britton, R. B. Blakestad, J. D. Jost, C. Langer, R. Ozeri, S. Seidelin, and D. J. Wineland, *Experimental purification of two-atom entanglement*, Nature 443, 838 (2006).

[Rei06b] R. Reichle, D. Leibfried, R. B. Blakestad, J. Britton, J. D. Jost, E. Knill, C. Langer, R. Ozeri, S. Seidelin, and D. J. Wineland, *Transport dynamics of single ions in segmented microstructured Paul trap arrays*, Fortschr. d. Phys. 54, 666 (2006).

[Ric73]   J. A. Richards, R. M. Huey, and J. Hiller, *A new operating mode for the quadrupole mass filter*, Int. J. Mass Spectrom. Ion Phys. 12, 317 (1973).

[Ric75]   J. A. Richards, *On the choice of steps in the piecewise-constant hill equation model of a quadrupole mass filter*, Int. J. Mass Spectrom. Ion Phys. 18, 11 (1975).

[Rie04]   M. Riebe, H. Häffner, C. F. Roos, W. Hänsel, J. Benhelm, G. P. T. Lancaster, T. W. Körber, C. Becher, F. Schmidt-Kaler, D. F. V. James and R. Blatt, *Deterministic quantum teleportation with atoms*, Nature 429, 734 (2004).

[Roh01]   H. Rohde, S. T. Gulde, C. F. Roos, P. A. Barton, D. Leibfried, J. Eschner, F. Schmidt-Kaler, and R. Blatt, *Sympathetic ground-state cooling and coherent manipulation with two-ion crystals*, J. Opt. B 3, S34 (2001).

[Roo00]   C. Roos, *Controlling the quantum state of trapped ions*, Doctoral thesis, University of Innsbruck (2000).

[Roo04a] C. F. Roos, G. P. T. Lancaster, M. Riebe, H. Häffner, W. Hänsel, S. Gulde, C. Becher, J. Eschner, F. Schmidt-Kaler, and R. Blatt, *Bell states of atoms with ultralong lifetimes and their tomographic state analysis*, Phys. Rev. Lett. 92, 220402 (2004).

[Roo04b] C. F. Roos, M. Riebe, H. Häffner, W. Hänsel, J. Benhelm, G. P. T. Lancaster, C. Becher, F. Schmidt-Kaler, and R. Blatt, *Control and measurement of three-qubit entangled states*, Science 304, 1478 (2004).

[Roo06] C. F. Roos, M. Chwalla, K. Kim, M. Riebe and R. Blatt, '*Designer atoms' for quantum metrology*, Nature 443, 316 (2006).

[Roo99] C. Roos, T. Zeiger, H. Rohde, H. C. Nägerl, J. Eschner, D. Leibfried, F. Schmidt-Kaler, and R. Blatt, *Quantum state engineering on an optical transition and decoherence in a Paul trap*, Phys. Rev. Lett. 83, 4713 (1999).

[Ros07] T. Rosenband, P. O. Schmidt, D. B. Hume, W. M. Itano, T. M. Fortier, J. E. Stalnaker, K. Kim, S. A. Diddams, J. C. J. Koelemeij, J. C. Bergquist, and D. J. Wineland, *Observation of the $^1S_0 \rightarrow {}^3P_0$ clock transition in $^{27}Al^+$*, Phys. Rev. Lett. 98, 220801 (2007).

[Ros08] T. Rosenband, D. B. Hume, P. O. Schmidt, C. W. Chou, A. Brusch, L. Lorini, W. H. Oskay, R. E. Drullinger, T. M. Fortier, J. E. Stalnaker, S. A. Diddams, W. C. Swann, N. R. Newbury, W. M. Itano, D. J. Wineland, and J. C. Bergquist, *Frequency ratio of Al^+ and Hg^+ single-ion optical clocks; metrology at the 17th decimal place*, Science 28, 1808 (2008).

[Rot06] B. Roth, J. C. J. Koelemeij, H. Daerr, and S. Schiller, *Rovibrational spectroscopy of trapped molecular hydrogen ions at millikelvin temperatures*, Phys. Rev. A 74, 040501 (2006).

[Row02] M. A. Rowe, A. Ben-Kish, B. DeMarco, D. Leibfried, V. Meyer, J. Beall, J. Britton, J. Hughes, W. M. Itano, B. Jelenkovic, C. Langer, T. Rosenband, and D. J. Wineland, *Transport of quantum states and separation of ions in a dual rf ion trap*, Quantum Inf. Comput. 2, 257 (2002).

[Sac00] C. A. Sackett, D. Kielpinski, B. E. King, C. Langer, V. Meyer, C. J. Myatt, M. Rowe, Q. A. Turchette, W. M. Itano, D. J. Wineland, and C. Monroe, *Experimental entanglement of four particles*, Nature 404, 256 (2000).

[Sau86] T. Sauter, W. Neuhauser, R. Blatt, and P. E. Toschek, *Observation of quantum jumps*, Phys. Rev. Lett. 57, 1696 (1986).

[Sch03]   F. Schmidt-Kaler, H. Häffner, S. Gulde, M. Riebe, G. P. T. Lancaster, T. Deuschle, C. Becher, W. Hänsel, J. Eschner, C. F. Roos, and R. Blatt, *How to realize a universal quantum gate with trapped ions*, Appl. Phys. B 77, 789 (2003).

[Sch04a]   T. Schätz, M. D. Barrett, D. Leibfried, J. Chiaverini, J. Britton, W. M. Itano, J. D. Jost, C. Langer, and D. J. Wineland, *Quantum dense coding with atomic qubits*, Phys. Rev. Lett. 93, 040505 (2004).

[Sch04b]   D. Schrader, I. Dotsenko, M. Khudaverdyan, Y. Miroshnychenko, A. Rauschenbeutel, and D. Meschede, *Neutral atom quantum register*, Phys. Rev. Lett. 93, 150501 (2004).

[Sch05]   P. O. Schmidt, T. Rosenband, C. Langer, W. M. Itano, J. C. Bergquist, and D. J. Wineland, *Spectroscopy using quantum logic*, Science 309, 749 (2005).

[Sch06]   S. Schulz, U. Poschinger, K. Singer, and F. Schmidt-Kaler, *Optimization of segmented linear Paul traps and transport of stored particles*, Fortschr. Phys. 54, 648 (2006).

[Sch07]   S. Schiller, *Hydrogenlike highly charged ions for tests of the time independence of fundamental constants*, Phys. Rev. Lett. 98, 180801 (2007).

[Sei06]   S. Seidelin, J. Chiaverini, R. Reichle, J. J. Bollinger, D. Leibfried, J. Britton, J. H. Wesenberg, R. B. Blakestad, R. P. Epstein, D. B. Hume, W. M. Itano, J. D. Jost, C. Langer, R. Ozeri, N. Shiga, and D. J. Wineland, *Microfabricated surface-electrode ion trap for scalable quantum information processing*, Phys. Rev. Lett. 96, 253003 (2006).

[Sho95]   P. W. Shor, *Scheme for reducing decoherence in quantum computer memory*, Phys. Rev. A 52, 2493 (1995).

[Sho96]   P. Shor, *Fault-tolerant quantum computation*, 37th Ann. Symp. Comp. Science (FOCS'96), and arXiv:quantph/9605011 (1996).

[Skl02]   S. E. Sklarz and D. J. Tannor, *Loading a Bose-Einstein condensate onto an optical lattice: An application of optimal control theory to the nonlinear Schrödinger equation*, Phys. Rev. A 66, 053619 (2002).

[Sor99]   A. Sorensen and K. Molmer, *Quantum computation with ions in thermal motion*, Phys. Rev. Lett. 82, 1971 (1999).

[Ste00]   A. Steane, C. F. Roos, D. Stevens, A. Mundt, D. Leibfried, F. Schmidt-Kaler, and R. Blatt, *Speed of ion-trap quantum-information processors*, Phys. Rev. A 62, 042305 (2000).

[Ste06a]   A. M. Steane, *How to build a 300 bit, 1 Giga-operation quantum computer*, arXiv:quant-ph/0412165 (2006).

[Ste06b]   M. Steffen, M. Ansmann, R. C. Bialczak, N. Katz, E. Lucero, R. McDermott, M. Neeley, E. M. Weig, A. N. Cleland, J. M. Martinis, *Measurement of the entanglement of two superconducting qubits via state tomography*, Science 313, 1423 (2006).

[Ste86]   S. Stenholm, *The semiclassical theory of laser cooling*, Rev. Mod. Phys. 58, 699 (1986).

[Ste96]   A. M. Steane, *Error correcting codes in quantum theory*, Phys. Rev. Lett. 77, 793 (1996).

[Ste97]   A. Steane, *The ion trap quantum information processor*, Appl. Phys. B 64, 623 (1997).

[Ste98]   A. Steane, *Quantum computing*, Rep. Prog. Phys. 61, 117 (1998).

[Sti06]   D. Stick, W. K. Hensinger, S. Olmschenk, M. J. Madsen, K. Schwab, and C. Monroe, *Ion trap in a semiconductor chip*, Nature Physics 2, 36 (2006).

[Str55]   H. Straubel, *Zum Öltröpfchenversuch von Millikan*, Naturwiss. 18, 506 (1955).

[Sym98]   R. R. Syms, T. J. Tate, M. M. Ahmad, and S. Taylor, *Design of a microengineered electrostatic quadrupole lens*, IEEE Trans. Electron Devices 45, 2304 (1998).

[Tan06]   D. J. Tannor, *Introduction to quantum mechanics: A time-dependent perspective*, University Science Books, Sausalito (2006).

[Tha99]   G. Thalhammer, *Frequenzstabilisierung von Diodenlasern bei 850, 854 und 866nm mit Linienbreiten im Kilohertz-Bereich*, Diploma thesis, University of Innsbruck (1999).

[Tim06]   N. Timoney, V. Elman, W. Neuhauser, and C. Wunderlich, *Error-resistant single qubit gates with trappes ions*, arXiv:0612106 (2006).

[Tim08]   N. Timoney, V. Elman, S. Glaser, C. Weiss, M. Johanning, W. Neuhauser, and C. Wunderlich, *Error-resistant single-qubit gates with trapped ions*, Phys. Rev. A 77, 052334 (2008).

[Thu03]   M. Thual, P. Chanclou, O. Gautreau, L. Caledec, C. Guignard, and P. Besnard, *Appropiate micro-lens to improve coupling between laser diodes and singlemode fibres*, Electronics Letters, 39, 1504 (2003).

[Tru07]   M. Trupke, J. Goldwin, B. Darquie, G. Dutier, S. Eriksson, J. Ashmore, E. A. Hinds, *Atom detection and photon production in a scalable, open, optical microcavity*, Phys. Rev. Lett. 99, 063601 (2007).

[Tur00]   Q. A. Turchette, D. Kielpinski, B. E. King, D. Leibfried, D. M. Meekhof, C. J. Myatt, M. A. Rowe, C. A. Sackett, C. S. Wood, W. M. Itano, C. Monroe, and D. J. Wineland, *Heating of trapped ions from the quantum ground state*, Phys. Rev. A 61, 063418 (2000).

[Tur98]   Q. A. Turchette, C. S. Wood, B. E. King, C. J. Myatt, D. Leibfried, W. M. Itano, C. Monroe, and D. J. Wineland. *Deterministic entanglement of two trapped ions*. Phys. Rev. Lett. 81, 3631 (1998).

[Van01]   L. M. K. Vandersypen, M. Steffen, G. Breyta, C. S. Yannoni, M. H. Sherwood, and I. L. Chuang, *Experimental realization of Shor's quantum factoring algorithm using nuclear magnetic resonance*, Nature 414, 883 (2001).

[Van05]   L. M. K. Vandersypen and I. L. Chuang, *NMR techniques for quantum control and computation*, Rev. Mod. Phys. 76, 1037 (2005).

[Wan08]   S. X. Wang, J. Labaziewicz, Y. Ge, R. Shewmon, and I. L. Chuang, *Individual addressing of ions using magnetic field gradients in a surface-electrode trap*, arXiv:0811.2422 (2008).

[Wak92]   I. Waki, S. Kassner, G. Birkl, and H. Walther, *Observation of ordered structures of laser-cooled ions in a quadrupole storage ring*, Phys. Rev. Lett. 68, 2007 (1992).

[Wal05]   P. Walther, K. J. Resch, T. Rudolph, E. Schenck, H. Weinfurter, V. Vedral, M. Aspelmeyer and A. Zeilinger, *Experimental one-way quantum computing*, Nature 434, 169 (2005).

[Wal08]   M. Wallquist, P. Rabl, M. D. Lukin, and P. Zoller, *Theory of cavity-assisted microwave cooling of polar molecules*, New Journal of Physics 10, 063005 (2008).

[Wes08]   J. H. Wesenberg, *Electrostatics of surface-electrode ion traps*, Phys. Rev. A 78, 063410 (2008).

[Wil02]   G. Wilpers, T. Binnewies, C. Degenhardt, U. Sterr, J. Helmcke, and F. Riehle, *Optical Clock with Ultracold Neutral Atoms*, Phys. Rev. Lett. 89, 230801 (2002).

[Win03]   D. J. Wineland, M. Barett, J. Britton, J. Chiaverini, B. DeMarco, W. M. Itano, B. Jelenkovic, C. Langer, D. Leibfried, V. Meyer, T. Rosenband, and T. Schätz, *Quantum information processing with trapped ions*, Phil. Trans. Royal Soc. Lond. A 361, 1349 (2003).

[Win75]   D. J. Wineland and H. Dehmelt, *Proposed 10^{-14} laser fluorescence spectroscopy on $T1^+$ mono-ion oscillator*, Bull. Am. Phys. Soc. 20, 637 (1975).

[Win78]   D. J. Wineland, R. E. Drullinger, and F. L. Walls, *Radiation-pressure cooling of bound resonant absorbers*, Phys. Rev. Lett. 40, 1639 (1978).

[Win79]   D. J. Wineland and W. M. Itano, *Laser cooling of atoms*, Phys. Rev. A 20, 1521 (1979).

[Win87]   D. J. Wineland, W. M. Itano, J. C. Bergquist, R. G. Hulet, *Laser-cooling limits and single-ion spectroscopy*, Phys. Rev. A 36, 2220 (1987).

[Wu96]   Y. Wu, *Effective Raman theory for a three-level atom in the Λ configuration*, Phys. Rev. A 54, 1586 (1996).

[Wun07]   C. Wunderlich, T. Hannemann, T. K. Körber, H. Häffner, C. Roos, W. Hänsel, R. Blatt, and F. Schmidt-Kaler, *Robust state preparation of a single trapped ion by adiabatic passage*, J. Mod. Opt. 54, 1541 (2007).

[Ye01]   J. Ye, L. S. Ma, and J. L. Hall, *Molecular iodine clock*, Phys. Rev. Lett. 87, 270801 (2001).

[Ye08]   J. Ye, H. J. Kimble, and H. Katori, *Quantum state engineering and precision metrology using state-insensitive light traps*, Science 320, 1734 (2008).

[Yu91]   N. Yu, W. Nagourney, and H. Dehmelt, *Demonstration of new Paul-Straubel trap for trapping single ions*, J. Appl. Phys. 69, 3779 (1991).

[Yu95]   N. Yu and W. Nagourney, *Analysis of Paul-Straubel trap and its variations*, J. Appl. Phys. 77, 3623 (1995).

[Zve61]   A. I. Zverev and H. J. Blinchikoff, *Realization of a filter with helical components*, Proc. of the IRE 8, 99 (1961).

[Zzy]   The fully unexspected imperfect Doppler cooling was also observed by Roos [Roo00]. A change in the polarization in one of the two 397nm Doppler cooling beams has allowed the variation of the mean quantum number between 15 and 250. In this experiment the non-thermally vibrational state distribution is caused by the imperfect voltage supplies. It changed instantly to Doppler cooling in the Lamb-Dicke regime by using battery-supplied voltages. The interaction of the ion with the noisy environment was responsible for this effect, showing the necessity of low noise devices for scalable microtraps.

[Zzz]   The continuous operation of the oven generating the neutral calcium beam influences the experimental results of the quantum jump spectroscopy. Because of the limited background pressure of 10^{-10}mbar and the loss during shuttling experiments a permanent atomic beam is required. The current of 2.0A...3.5A for the oven effects lineshifts on the order of 6.0kHz/A on the carrier - in general more than the carrier linewidth. In spite of the active stabilized current supply for the magnetic field coils it is obvious, that the unstabilized current supply for the oven effectuates magnetic field noise. This can cause decoherence effects and dephasing of the Rabi oscillations. The heating rate measurements with a single ion are realized after 8 months of continuous operation of the oven. Therefore it can be assumed that patch potentials influences the measurements in this spatial region and the results can be interpreted as an upper limit for the trap heating rate.